THE GARBAGE PRIMER

The League of Women Voters Education Fund

LYONS & BURFORD, PUBLISHERS

Copyright © 1993 by The League of Women Voters Education Fund.

Design by M. R. P.

Printed in the United States of America on recycled stock that includes 15% post-
consumer waste.

10 9 8 7 6 5 4 3 2 1

Library of Congress Cataloging-in-Publication Data

Murphy, Pamela.
    The garbage primer / the League of Women Voters Education Fund;
[researched and written by Pamela Murphy with assistance from
Christine Mueller and Mamatha Gowda].
        p. cm.
    Includes bibliographical references and index.
    ISBN 1-55821-250-7 :
    1. Refuse and refuse disposal—United States. 2. Refuse and
refuse disposal—United States—Decision making—Citizen
participation.   I. Mueller, Christine (Christine R.)   II. Gowda,
Mamatha.   III. League of Women Voters (U.S.).   Education Fund.
IV. Title.
TD788.M87   1993
363.72'8—dc20   93-20639
                                                              CIP

ACKNOWLEDGMENTS

The *Garbage Waste Primer* was researched and written by Pamela Murphy with assistance from Christine Mueller and Mamatha Gowda; it was edited by Monica Sullivan. The project was directed by Elizabeth Kraft. The LWVEF wishes to thank the project's Advisory Committee and the many other reviewers of the draft manuscript. *The Garbage Primer* was made possible by grants from the U.S. Environmental Protection Agency and an anonymous donor.

LWVEF Chair: Becky Cain

Vice-Chairs: Diane Sheridan and Peggy Lucas

Project Trustee: Nancy Pearson

Executive Director: Gracia Hillman

Director: Sherry Rockey

Natural Resources Program Manager: Elizabeth Kraft

This book is available from bookstores, from the publisher, and from the League of Women Voters of the United States, 1730 M Street NW, Washington DC 20036, (202) 429-1965. Pub. #954.

CONTENTS

INTRODUCTION

Today, taking out the trash is everyone's responsibility. Public concern over the environmental and economic impacts of what we throw away has prompted local, state, and federal governments to rethink municipal solid waste management strategies. To tackle their growing piles of trash, many communities are integrating management options—source reduction, recycling, composting, incineration, and landfilling. This citizen's handbook is intended to help citizens and local officials better understand the components of an integrated municipal solid waste management system and to develop criteria to evaluate the different options available. It is *not* intended to be a step-by-step guide for managing garbage because every community must custom-design its own plan.

As you become involved in solid waste issues, you may be overwhelmed by the accumulation of facts and statistics. One can find information and studies to support just about any view. But, take heart. To understand the issues facing your community, you should evaluate the differing opinions and take action; you do not need scientific or technical expertise, just a willingness to learn. Citizens can readily understand and influence municipal solid waste issues and policies.

The handbook outlines the technical, environmental,

economic, and policy issues related to each waste management option, and discusses the management of special wastes and the siting and financing of municipal solid waste facilities. To cover the gamut of viewpoints on how best to handle our garbage, topical issues and local programs are presented in boxes throughout the text. The concluding chapter, "What Citizens Can Do," provides guidelines to help evaluate waste management options, and includes suggestions for taking action on municipal solid waste issues in your community.

ONE

AFTER THE TRASH CAN

As did generations before us, we continue to search for solutions to our garbage problem once the trash can is full. The dilemma is not new; only today our waste management options are more sophisticated and the waste we generate is more diverse and plentiful.

Garbage, trash, waste, rubbish—all are familiar terms for what is technically called municipal solid waste. The U.S. Environmental Protection Agency (EPA) defines municipal solid waste, or MSW, as "wastes such as durable goods, nondurable goods, containers and packaging, food scraps, yard trimmings, and miscellaneous inorganic wastes from residential, commercial, institutional, and industrial sources." According to EPA, municipal solid waste does *not* include "wastes from other sources, such as construction and demolition wastes, municipal sludges, combustion ash, and industrial process wastes that might also be disposed of in municipal waste landfills or incinerators."[1] Many states and municipalities use a broader definition of municipal solid waste than EPA's, often including construction wastes and automobile scraps, for example.

A "throwaway" mentality, combined with a burgeoning environmental awareness, leaves us with a growing pile of

| 1960 | 1970 | 1980 | 1990 | 2000 |

Figure 1. Per capita generation of municipal solid waste, 1960 to 2000, in pounds per person per day. This graph shows waste generation before materials recovery or incineration. Demographic changes, economic factors, and consumer preferences are among the factors contributing to the increase in per-capita generation of MSW. *Source:* Adapted from EPA, *Characterization of Municipal Solid Waste in the United States: 1992 Update.*

trash and fewer acceptable ways to dispose of it. The amount of garbage we throw out in this country has more than doubled over the past 30 years, while the population has increased by only 38 percent. In 1960, Americans threw out 88 million tons of garbage and by 1990 we were filling our trash cans with 195.7 million tons of waste. If we do not change our ways, we will usher in the 21st century with more than 222 million tons of garbage a year. The greatest increases are in nondurable items such as paper and clothing, durable items such as appliances and tires, and containers and packaging (see Figures 1 and 2).

Managing this growing heap of trash is primarily the responsibility of local governments. As local taxpayers, we spend more than $30 billion every year to have our trash taken care of. To alleviate the burden on local governments already challenged by tight budgets, solid waste officials are integrating solid waste management options—source reduction, recycling, composting, incineration, and landfilling—to form comprehensive solid waste management plans.

In 1990, 67 percent of our garbage was buried in the coun-

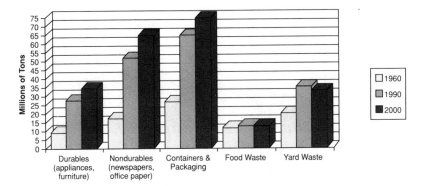

Figure 2. Types of products generated in the waste stream, 1960 to 2000, in millions of tons. This graph represents generation before materials recovery or combustion; it does not include construction and demolition debris or industrial process wastes. *Source:* Adapted from EPA, *Characterization of Municipal Solid Waste in the United States: 1992 Update.*

try's 6,000 landfills, 16 percent was burned in 176 incinerators and waste-to-energy plants (incinerators generating energy by burning garbage), and 17 percent was recycled and composted. By 1995, EPA estimates that 58 percent of our garbage will be sent to landfills, 17 percent burned, and 25 percent recycled and composted. But, despite this shift away from our historical dependence on landfills toward recycling and incineration, landfills will continue to be needed for the disposal of nonrecyclables and incinerator ash. The trend in municipal waste management from 1960 to 2000 is shown in Figure 3. After the 1960s, the amount of garbage incinerated dropped significantly, as old incinerators were closed down due to air pollution regulations and then gradually increased as the waste-to-energy industry grew. The amount recycled steadily increased, while the amount landfilled peaked in the mid-1980s.

In a country with an abundance of open space, the issue is not that we are running out of land for landfills, but that we are running out of socially and politically acceptable sites to put them. Federal, state, and local rules for siting and operating waste facilities, particularly landfills, are growing more stringent. As of 1993, new

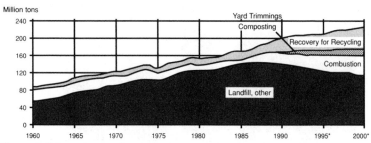

Figure 3. Management of municipal solid waste in the United States, 1960 to 2000. An integrated waste management approach with a focus on source reduction, recycling, and composting can reduce the MSW generated and increase recovery for recycling. *Estimated. *Source:* EPA, *Characterization of Municipal Solid Waste in the United States: 1992 Update.*

EPA regulations require landfill owners and operators to comply with a minimum set of standards. In anticipation of the new regulations some old landfills have closed, and the costs to build new landfills are expected to increase. The growing awareness of the health and environmental risks posed by heavy metals and other pollutants also contributes to strong public opposition to siting new waste facilities. Figure 4 lists potential health and environmental risks of certain pollutants.

In 1976, the federal government enacted the Resource Conservation and Recovery Act (RCRA) as the principal national law regulating the management of solid waste. RCRA is designed to encourage communities to manage solid and hazardous wastes in an environmentally sound manner that maximizes the use of valuable resources, including energy and materials that are recoverable from solid waste. Under Subtitle D of RCRA, the federal government set criteria and minimum technical requirements for environmentally acceptable municipal solid waste disposal facilities, and states are responsible for enforcement. By design, RCRA does not provide for a comprehensive federal regulatory role in municipal solid waste management; thus states have passed state regulations for landfills and incinerators and have set recycling rate goals.

Since RCRA is to be considered for reauthorization in Congress, new efforts will be made to direct funding and federal regulatory authority toward municipal solid waste. The political debate on the federal government's role in solid waste management is taking form over a variety of issues, including recycling requirements, incinerator restrictions, and a national bottle bill.

500 B.C.

Athens organizes the first municipal dump in the western world. Waste was required to be disposed of at least one mile from city walls.

Underlying this current debate on solid waste management is the more fundamental debate of how we should pay for waste management. Some hold the consumer ultimately responsible and believe that, as taxpayers, we should pay the local government to dispose of the waste we create. Others believe that industry should be held responsible for the waste it generates, and that waste management costs should be built into product costs and paid by the product producers and consumers.

To address the economic and environmental issues surrounding solid waste management, communities increasingly are relying on a combination of techniques to handle their garbage. This handbook presents an integrated solid waste management approach that encourages communities to custom-design a package of waste management practices to complement their demographic, geographic, and waste characteristics, and the availability of recycling markets. The waste management options presented in the handbook are arranged according to EPA's solid waste management "hierarchy"—source reduction, recycling and composting, incineration, and landfill. EPA adopted this hierarchy to minimize the country's dependence on landfills and, in doing so, gave priority to source reduction and recycling over disposal-based approaches. Although this hierarchy is somewhat controversial in the field of waste management, it provides a useful rule of thumb to follow when evaluating the different components of an integrated approach to waste management.

Figure 4. Some Pollutants Found in the Municipal Waste Stream

Pollutant	Major Health Effects	Some MSW Sources	Comments
Arsenic	Carcinogen. Can cause skin cancer and when inhaled, associated with lung cancer; affects intestines and liver.	Household pesticides, wood preservatives.	Popularly known for its use as a poison in murder mysteries. It is a naturally occurring element used in different household products.
Cadmium	Carcinogen. Damages kidneys; when inhaled, associated with lung cancer.	Nickel-cadmium batteries (used in consumer goods), plastics (used as a stabilizer in polyvinyl chloride (PVC) plastics and as a plastic pigment), paints, enamel pigments, colored printing inks for packaging.	Nickel-cadmium batteries (rechargeable). Batteries account for an estimated 54% of cadmium in the waste stream after recycling.
Chromium	Causes kidney, liver, and nervous and circulatory system damage; respiratory problems; when inhaled, associated with lung cancer.	Pigments for colored papers and paper coating, paints, wood preservatives, batteries: alkaline, lithium cell and zinc-carbon.	A blue-white solid found in rocks and minerals.
Dioxins	Probable carcinogen. Causes liver and kidney damage, gastric ulcers, reproductive and developmental problems.	No natural source. A by-product formed when chlorine-containing products are incompletely burned (e.g., paper, PVC plastics).	EPA is conducting a review of the health risks posed by dioxin, which has been considered a probable carcinogen for many years. The preliminary findings suggest that dioxin may cause damage to immune and reproductive systems and birth defects at very low levels of exposure

Hydrochloric Acid	Causes respiratory problems and eye irritation.	Paper and PVC plastics.	Hydrochloric acid contributes to the formation of acid rain.
Lead	For children can cause mental retardation and learning disabilities; damages kidneys, liver, brain nerves, and heart; affects blood pressure. Especially toxic to children and pregnant women.	Lead-acid car batteries, electrical equipment, plastics (used as a stabilizer in PVC plastics and as a plastic pigment), colored printing inks for packaging, paints, insecticides.	Although lead-based paints were banned for residential use in 1978, 57 million American homes built before 1980 still contain some lead paint.
Mercury	Causes neurological and kidney damage, blindness; associated with birth defects. Especially toxic to aquatic life.	Most types of consumer batteries, paints, electrical equipment, fluorescent lights, plastics, dyes, thermometers.	Household batteries account for 88% of the total mercury content of municipal solid waste. The use of mercury by U.S. manufacturers declined 33% from 1984 to 1989, primarily due to the virtual elimination of mercury from alkaline batteries.
Methane	Explosive and can cause asphyxiation.	Decomposition of waste in landfills.	Methane is a greenhouse gas.

Sources: Environmental Defense Fund, *Recycling & Incineration: Evaluating the Choices*, INFORM, *Making Less Garbage: A Planning Guide for Communities.*

TWO

SOURCE REDUCTION

No approach to municipal solid waste management is more widely praised and less widely practiced than the simple idea of producing less garbage in the first place. As a concept, source reduction is given top billing and sits at the top of the Environmental Protection Agency's waste management hierarchy. In practice, it is still unclear where source reduction starts. Local governments are calling for state action, states are calling for federal initiatives, consumers are demanding action from manufacturers, and manufacturers are urging individuals to change their behavior. The concept of source reduction—to decrease the volume and toxicity of garbage—may be a simple one, but getting from theory to practice is proving to be a difficult task.

To practice source reduction, consumers, government, industries, and businesses can all change present practices by:

- Decreasing consumption;
- Reusing products;
- Using fewer materials;
- Using less toxic materials;
- Making and using longer-lasting products.

1388
_____

▼

English Parliament bans waste
disposal in public waterways and
ditches.
_____

The benefits of a source reduction program are straightforward. Source reduction reduces the cost of waste management because less waste needs to be collected, processed, and disposed, decreases the negative environmental impacts and potential health risks that must be mitigated, and requires the use of fewer materials and natural resources. Why then are source reduction efforts not more widely practiced?

There are several reasons. First, source reduction is a nontraditional approach to solid waste management. It shifts the focus of waste management from managing waste after it is thrown out to preventing waste from being generated. Second, source reduction depends on a fundamental change in the consumptive- and convenience-oriented behavior and attitudes of consumers, manufacturers, and policy makers. Some say this behavioral change would place an unnecessary burden on consumers and manufacturers. And finally, the economic incentives are not as strong as they could be to encourage manufacturers to change production processes or consumers to purchase fewer products or less toxic products. For example, manufacturers are not charged for the garbage their products generate; thus, product prices do not reflect the environmental costs associated with their manufacture, consumer use, and disposal. And, consumers typically are not charged a higher price for purchasing or disposing of toxic products.

The good news is that strategies to achieve source reduction of municipal solid waste *are* being developed. While manufacturers always have sought ways to reduce their use of materials as a measure to keep production costs low, new efforts by them, as well as by consumers and state and local governments, are demonstrating that source reduction is an effective waste management option. This chapter focuses on a few actions government, industry, and consumers can take to practice source reduction.

THE CONFUSION OVER WEIGHT AND VOLUME

Garbage is measured by weight and volume. Of the two, weight figures are the most commonly used because of the ease of calculating a pound or ton of trash. The weight of garbage often determines the price of recyclables and landfill and incinerator charges. Volume figures are useful to determine landfill capacity, collection vehicle capacity, and the feasibility of reducing the quantity of materials.

Source reduction strategies also factor in weight and volume considerations. Reducing the weight of a package or product is a large component of source reduction. But confusion can arise when it is assumed that a lighter product will be smaller in size. For example, the amount of aluminum or plastic used in a container may be reduced, but the container may still take up the same amount of space in the landfill. To be effective, source reduction efforts should reduce product/packaging weight *without* increasing volume, or reduce volume *without* increasing weight. One sure way to accomplish this is to focus on eliminating unnecessary packaging layers.

WHAT GOVERNMENT CAN DO

State and local governments play a critical role in source reduction. They have the authority to set source reduction goals and the ability to promote its practice by citizens, businesses, and industry.

The role that the federal government can play in source reduction programs may be less obvious, but it is important. Source reduction practices are greatly influenced by national and international commerce; therefore, the federal government can promote

1400

Garbage piles up so high outside Paris gates that it interferes with the city's defenses.

source reduction by establishing tax incentives for reduction, developing procurement guidelines for source-reduced products, or researching the measurement and monitoring of source reduction. Government agencies also can play the much needed role of educator. By educating citizens nationwide on ways to practice source reduction, the federal government can support state and local government efforts to reduce the amount of waste being generated.

Legislation and regulation

Legislation and regulation to promote source reduction is gaining acceptance, with states and municipal governments instituting a variety of strategies to minimize the amount of garbage going into the waste stream.

Solid Waste Plans. State and local solid waste plans provide a framework for managing garbage over a period of time, generally 5–10 years, and set specific waste management goals, such as recycling a certain percentage of the waste stream by a certain date. By including source reduction goals in state plans, as Michigan has done, states can encourage source reduction practices at the local level. In addition to state waste plans, some states and communities require businesses to develop their own source reduction plans. For example, the state of Rhode Island requires businesses employing more than 100 persons to submit waste plans to the state that include source reduction programs.

Packaging Legislation. Federal legislation could be the most effective way to reduce the amount of packaging used in the United States because product marketing and distribution tends to be national in scope. In 1992, packaging legislation (also called "rates and dates") under the National Waste Reduction and Management Act passed the full House Energy and ded byCommerce Committee. This legislation would require packagers to use only glass, aluminum, steel, bimetal, or plastic packages that meet at least one of five requirements: an industry-wide recovery rate, a company-specific recovery rate, a minimum recycled content requirement, packaging reuse for its original purpose, or a source reduction requirement. Another packaging initiative introduced in Congress in 1992 called for reducing the use of four heavy metals—lead, cadmium, mercury, and hexavalent chromium—in packaging. The proposal, modeled after legislation developed by the Coalition of Northeast Governors (CONEG), would have prohibited the sale of packaging containing these metals. Although the U.S. Congress did not pass this proposal, 14 states (Connecticut, Georgia, Illinois, Iowa, Maine, Maryland, Minnesota, New Hampshire, New Jersey, New York, Rhode Island, Vermont, Washington, and Wisconsin) have adopted legislation based on CONEG's model bill. Similar packaging legislation may be reintroduced in Congress.

> **1690**
>
> First paper mill in U.S. established at the Rittenhouse Mill near Philadelphia; it is essentially a recycling industry with paper made from recycled fibers (waste paper and old rags).

Canada is one of several countries to pass national packaging legislation. The National Packaging Protocol relies on voluntary compliance by manufacturers to meet a national goal of 50 percent reduction in the amount of waste sent to landfills and incinerators by the year 2000. (See *Canada Targets Packaging,* pp. 18–19.)

State legislatures across the United States also are debating packaging legislation. Many of the bills are modeled after a proposal developed by the Massachusetts Public Interest Research Group (MassPIRG), which combines source reduction and recycling goals. The model bill is designed to reduce the amount of packaging requiring disposal, but also allows recycling to be substituted for source reduction.

1800

▼

Patent for paper using deinked waste paper as part of its fiber source is issued in London.

Voluntary source reduction programs are another option. An example is the "CONEG Challenge" issued to major industries in the United States to voluntarily reduce their excess packaging. A set of "Preferred Packaging Guidelines" suggests steps industry can take to eliminate, minimize, reuse/refill, and recycle packaging. By the end of 1992, 35 companies had accepted CONEG's challenge and agreed to set packaging reduction goals and report their progress to CONEG.

Environmental Labeling. Labeling policies also can encourage source reduction. Such policies require manufacturers to meet minimum product standards before making such product claims as "refillable," "less packaging," "ozone safe," "recyclable," "compostable," and "degradable." The Federal Trade Commission has developed voluntary guidelines for environmental marketing claims on product labels and in advertisements. The guidelines are designed not to define environmental labeling terms conclusively, but to identify the types of environmental claims that should be explained or qualified to avoid deceiving consumers. These guidelines are not legally enforceable and do not preempt state and local laws. By 1992, 13 states had adopted their own legislation to regulate or to authorize the regulation of environmental labeling on products and packaging. (For further discussion on "green" labeling, see *The Many Shades of Green,* pp. 21–24).

Product Design. Regulating the design of products is another policy option. Under this type of regulation, products that do not meet certain design criteria can be taxed or have their use restricted. Design criteria also can require the extension of product warranties to increase the life of a product and encourage product durability, or they can require the reduction of hazardous components in a product. For example, California, Connecticut, Minnesota, New

York, Oregon, and Vermont have passed laws requiring that all batteries be mercury-free by 1996, and battery manufacturers have responded by developing batteries without this toxic component.

1842

A report in England links disease to filthy environmental conditions, and helps launch the "age of sanitation."

Bans. Some state and local governments have instituted bans on the use or disposal of certain items to reduce their quantity in the waste stream. These bans may be based on the presence of toxic materials in a product, such as mercury in household batteries; on the volume of a particular material in the waste stream, such as yard wastes; the wasteful use of resources, such as multi-material packaging. Bans are a direct policy approach, but they are controversial because of the legal ramifications surrounding interstate, regional, and local commerce laws.

Proponents believe that bans accomplish several goals, including decreasing the volume of the banned material being sent to landfills and incinerators, prompting industry to change production practices, increasing public awareness, and directly addressing environmental and public health concerns. Opponents argue that some bans are a symbolic backlash against our perceived "throwaway society" and they accomplish little, if anything at all. They assert that bans simply force substitution of one material for another and do not increase the demand for recyclable materials, but rather eliminate the option to recycle the banned material while depriving consumers of the desirable features of that material.

To evaluate the appropriateness of a ban, a community should consider the intended goal and the possible unintended economic and environmental consequences. The most common types of bans used to promote source reduction are product or packaging bans. Product bans target specific products or contaminants in products. They are designed to force the use of more environmentally benign products or the removal of potential contaminants from the waste stream. For example, Maine has banned aseptic packaging (juice boxes). Berkeley, California, and Portland, Oregon, are among some of the cities that have banned polystyrene food packaging (e.g., fast-food clamshells).

Product bans are controversial because the trade-offs between banned and replacement products are often not clear-cut. For example, banning polystyrene foam cups may remove a product from the waste stream that many consider harmful to the environment, but the possible substitutes may create their own negative impacts. It is necessary to compare the impacts over the life of the possible substitute products to those of the banned product. If paper cups are to replace polystyrene foam cups, then the impacts of production, transportation, and disposal of both products should be compared. (See Figure 5 for more on what government can do to encourage source reduction.)

CANADA TARGETS PACKAGING

To counter the rising level of packaging in its waste stream, the Canadian Council of Ministers commissioned the National Task Force on Packaging to develop a national strategy to reduce Canada's packaging waste 50 percent by the year 2000. The National Task Force included representatives from federal, provincial/territorial and municipal governments, industry, and environmental and consumer groups. The result was the National Packaging Protocol, adopted in March 1990.

Under the Protocol the goals established for reducing packaging in the waste stream are 20 percent from 1988 levels by December 1992, 35 percent by December 1996, and 50 percent by December 2000. The reductions are to be achieved through source reduction, reuse, and recycling. Incineration is not considered a reduction option. While the Protocol allows manufacturers to voluntarily reduce packaging at specified target levels, a

continued on next page

regulatory framework will be implemented if packaging reductions are not achieved.

In an effort to minimize the impact that packaging has upon the environment, the Task Force also called for the development of guidelines for conducting environmental "profiles," commonly referred to as life-cycle analysis. The life-cycle analysis of packaging materials will identify environmental impacts generated through the manufacture, use, and post-use management of packaging. The goal of this undertaking is not only to allow consumers to compare packaging options, but to prompt manufacturers to implement steps to achieve reductions in packaging.

The Task Force plans to develop a campaign to educate all Canadians about the environmental impact of packaging and to encourage informed purchasing practices. The Task Force also has prepared a technical database on the management of packaging and commissioned technical reports on packaging.

Economic incentives and disincentives

Government also can use financial incentives, such as tax credits and subsidies to encourage source reduction, or use financial disincentives, such as taxes and fees to discourage waste-producing activities. Economic policies targeted at either industry or consumers can "correct" product prices so that they reflect their environmental costs (or "externalities"). This approach allows industry and consumers to choose between paying to practice business as usual or saving money by changing current practices.

Variable Can Fees. To encourage waste reduction many communities are implementing a waste collection system that charges residents and businesses according to the *amount* of garbage placed at the curb instead of charging a flat fee. This type of fee structure

---
1850

▼

Paper is manufactured from wood.

---

goes by a variety of names—including pay-per-can, pay-as-you-throw, volume- or weight-based pricing, pay-by-the-bag, and unit-based pricing.

The pay-per-can garbage collection system has long been a component of many west coast communities. For example, San Francisco has been charging for garbage services this way since 1932, and Olympia, Washington, since 1954; similar programs are only now gaining popularity in communities in other parts of the country. Today, more than 200 communities in 19 states use pay-per-can fees. Some communities such as Seattle, Washington, Durham, North Carolina, and Farmington, Minnesota, are taking this concept one step further and testing the feasibility of a weight-based pricing system. (See *Pay-As-You-Throw*, p. 49).

The primary interest of state and local governments implementing a variable can fee is to encourage source reduction, recycling, and backyard composting. Other advantages of this system are that trash fees can more realistically reflect the true cost of solid waste management in the community, and that fees are more equitable. An early criticism that variable fees would encourage illegal dumping and contamination of curbside recycling bins has proven to be unfounded or only a temporary problem in the more than 100 communities surveyed using this fee structure.[2]

Subsidies. Another way government can encourage participation in a source reduction program is by subsidizing both industries that use source reduction techniques and consumers who purchase reusable, durable, and repairable products. Subsidies can encourage source reduction activities that might not otherwise happen because of inadequate market prices for source reduced products or other disincentives. Subsidies can help to support industrial activities such as product redesign, research and development, and the purchase of new equipment. Although industries are not currently subsidized, government could provide tax credits, loans, or or grants for source reduction pilot projects or efforts to standardize products.

Taxes/Fees. While subsidies can be used to encourage source reduction, taxes can act as a disincentive for practices that create high environmen-

> **1874**
>
> ▼
>
> In Nottingham, England, a new technology called "the destructor" provides the first systemic incineration of municipal refuse.

tal costs. Taxes can discourage the use of certain products, such as seldom-recycled products or packaging. An advance disposal fee (ADF) is one type of tax that can be levied by states. This nonrefundable fee can be levied on either the manufacturer, wholesaler, retailer, or consumer of a product. Revenues generated from the fee are used to subsidize or offset recycling or disposal costs of the taxed product. Although states are discussing the merits of ADFs, not one has enacted a true ADF. Instead, states have passed a similar type of tax, a surcharge paid by consumers on hard-to-dispose-of items such as tires, motor oil, lead-acid batteries, major appliances, cars, and antifreeze. The theory is that individuals, rather than society at large, should pay for the disposal of these special wastes. Consumers pay this tax when they purchase or when they dispose of certain products. Tires, for example, are taxed in 18 states; consumers either pay the fee when they purchase tires or when they register or sell their car. (For further discussion on special wastes, see Chapter 6.)

THE MANY SHADES OF GREEN

In survey after survey consumers have shown their enthusiasm for "environmentally friendly" products, and manufacturers have been quick to respond. In grocery stores we are presented with an array of "green" or "environmentally friendly" products ranging from paper prod-
continued on next page

ucts made from recycled paper, recyclable plastic shopping bags, and nontoxic cleaners to concentrated
detergents and refillable bottles. The number of new green
products being introduced to the marketplace is stagger-
ing. According to EPA, in 1989 green products repre-
sented 4.5 percent of the total number of new products. In
1991, such products accounted for 13.4 percent of the new
products introduced.

The rush to tap the green market has led some manufacturers to make questionable environmental claims. As
consumers we must muddle through an assortment of
products, relying on the marketing departments of product
manufacturers to steer us toward the green products. The
Federal Trade Commission (FTC), quick to pull products
with blatantly false environmental claims from store
shelves, has developed national guidelines to help businesses make environmental claims and to help consumers
assess product claims. In July of 1992 the FTC issued
guidelines " . . . to protect consumers and to bolster their
confidence in environmental claims, and to reduce manufacturers' uncertainty about which claims might lead to
the FTC law-enforcement actions . . . " But some state
governments, organizations, and stores did not wait for
federal government action and forged ahead to establish
their own labeling requirements or programs.

The two national independent labeling efforts under
way are the Environmental Certification Program and
Green Seal. The Environmental Certification Program
(formerly the Green Cross program), a Scientific Certification Systems, Inc. program initiated in 1990, has
issued "green cross" emblems to more than 150 companies
for more than 500 products. The program has two types of
certification: a "claims certification" ensures the accuracy
of a product's environmental claim, such as recycled content or biodegradability, and a "product certification" as-

continued on next page

sesses a product's impact on the environment by evaluating the energy used, amount of air and water pollution, and the solid waste produced during the life of the product. This information is compiled onto an environmental report card that appears on the product's packaging. This approach is based on a form of life-cycle analysis, a methodology that takes into consideration the environmental burdens associated with each stage of a product's manufacture, distribution, use, and disposal.

Green Seal, a nonprofit environmental program, also launched in 1990, sets product standards that companies must meet to be eligible for a seal of approval. Products receive a stamp of approval if they meet program standards set through consultation with business, government, academia, and public interest groups. Green Seal has issued product standards for bathroom and facial tissue, printing and writing paper, compact fluorescent lamps, rerefined engine oil, and water-efficient fixtures.

Despite these programs, state laws, and the guidelines developed by the FTC, consumers still face many unanswered questions. What if a product meets the criteria for a green label but the packaging does not? Or what if the packaging is "green" but the product is not? For example, a roll of paper towels is made of recycled paper, but the plastic wrapper is made of virgin material; or the bottle for bleach is made of recycled plastic, but the product itself may be harmful to the environment. Do these products warrant a green label?

Is a company eligible for a green label program if it produces one or two products that meet labeling requirements but has a company performance record as a polluter? How do we assess an environmental claim when a company substitutes environmental action for a product that is environmentally "unfriendly?" For example, what if a company promotes the sale of a household cleaner or

continued on next page

> disinfectant that contains toxic chemicals as a green prod-
> uct in exchange for providing an environmental service
> such as planting a tree for every proof-of-purchase
> returned?
>
> In the end, what it comes down to is that a healthier
> environment means consuming *less* as well as consuming
> "greener."

WHAT INDUSTRY CAN DO

Industry and business have a pivotal role to play in promoting
source reduction at the local, state, and national levels. The commercial sector alone generates approximately 40 percent of the
country's municipal solid waste. Many companies have taken initiatives to decrease the toxicity and volume of their products, as
well as to improve operating efficiency and cut costs.

Product Design. Changes in the design of a product or its packaging, such as making the packaging lighter in weight or the product
smaller or in a concentrated form, can save transportation and
packaging costs. Aluminum cans are a good example of light-
weighting packaging. Over time, manufacturers of beer and soda
cans have decreased the amount of aluminum used per can. In
1976, 23 cans were made from a pound of aluminum; today, 29
cans are made from the same amount. Initiatives have been taken
by other manufacturers. For example, General Mills reduced the
thickness of the plastic bags in cereal boxes to decrease the amount
of plastic the company used annually by 500,000 pounds. Procter
& Gamble eliminated the packaging for Secret and Sure deodor-
ants, thereby removing about 80 million cartons a year from the
waste stream.

New product designs also can increase product durability and
therefore reduce waste by making products last longer. Radial tires,

for example, are more durable than earlier tire models because the tires are made with natural and synthetic rubbers and are strengthened by steel belts in the tires. Unfortunately, this improvement in the durability of tires has made them a challenge to recycle.

1885

▼

The first garbage incinerator in the U.S. is built on Governor's Island, New York.

Changes in product designs also can eliminate toxic chemicals from a product or its packaging. Apple Computer, Inc., for example, switched to brown (kraft) cardboard boxes from bleached white boxes.

Business Operations. Source reduction also can be achieved by changing daily business operations. Offices with high reproduction budgets may consider establishing a double-sided copying policy. After instituting such a policy for client documentation at its reproduction facilities, AT&T estimated that if this policy is followed only 50 percent of the time, the amount of paper used will be reduced by 77 million sheets annually and company costs will decrease by $385,000 a year.

Other practices that businesses can use are electronic mail, circulation of one document or memo (rather than multiple copies), and reuse or return of corrugated boxes and shipping pallets. Pepsi Cola replaced the corrugated cardboard boxes for shipping 2-liter bottles with reusable plastic crates and reduced the amount of cardboard the company uses each year by 80,000 tons.

Refillable bottles and containers offer another source reduction opportunity to industry. In the early 1960s, 89 percent of all soft drinks and 50 percent of all beer in this country was sold in refillable bottles. Today, the figure for refillable bottles hovers around 6 percent for soft drinks and beer, combined. In 1990, Rainer Brewing Company, a Seattle-based brewery, returned to the refillable bottle. In the first year, the company bought back and refilled 20 million bottles—enough bottles to fill a 60,000-plus seat stadium with bottles three feet deep and save enough energy to serve 1,434 homes for a year.

LIFE-CYCLE ANALYSIS: WHERE'S THE SCIENCE?

In a marketplace that is currently awash with "green products," are consumers on their own when it comes to evaluating the environmental soundness of a product, or can they turn to "life-cycle analysis" for some guidance? Life-cycle analyses, also referred to as "cradle to grave" studies and resource or environmental profiles, typically compare competing products by attempting to identify energy use, material inputs, and wastes generated during a product's life. The life-cycle of a product includes the extraction and processing of raw materials to make it, the manufacturing and transportation of the product to the marketplace, and finally, the use and disposal of the product. Life-cycle analysis is used to compare specific products, as well as to compare generic materials, such as plastic versus paper.

Manufacturers have used this type of analysis for marketing decisions and other internal decision-making activities for the past 20 years. The current generation of life-cycle analyses, however, has been extended into the public arena for use in consumer marketing and product labeling. This trend has prompted considerable debate over the overall concept of life-cycle analysis.

Based on discussions at a 1990 workshop of the Society for Environmental Toxicology and Chemistry (SETAC), a scientific and professional society, a model for life-cycle analysis was developed. The model consists of three components: inventory analysis, impact analysis, and improvement analysis.

The first component, an inventory of materials and energy used and environmental releases to air, water, and

continued on next page

land from all stages in the life of a product, is most widely used. The second component, an analysis of potential environmental impacts related to the use of energy and material resources and environmental releases, is not as widely used. Due mainly to the inherent complexity of the analysis required, it is difficult to determine the relative impacts of pollutants, especially when comparing environmental impacts of different materials and production processes. There may never be agreement on the relative impact of different pollutants. The third component, an analysis of the changes needed to help bring about environmental improvements for a product or process under study, draws upon the analysis of the other two components. The ultimate objective of life-cycle studies is to bring about actual improvements in products and processes.

In comparison to the SETAC model, current life-cycle analysis is at a very primitive stage. Findings often are based on assumptions that are subjective or even misleading, especially when studies use only the inventory type of analysis. For example, one product may claim that it produces fewer pollutants during manufacture, but those pollutants may, in fact, be far more toxic than those produced by its competition. Despite these shortcomings, many manufacturers use life-cycle analysis as a prime public relations tool to promote their products.

WHAT CONSUMERS CAN DO

Consumers can practice source reduction at home and at work, and they can use their purchasing power to support source reduction initiatives by industry. The following are three activities we can do:

Pre-cycling. By "pre-cycling" consumers can put their buying power to work for the environment. Consumers can decrease the amount of garbage thrown out and bolster the production of certain products and packaging by purchasing products and packaging that are recyclable, that are made from recycled material or from fewer materials, and that are less toxic than comparable products. Simply by shopping selectively—buying in bulk, selecting concentrated products and products with reduced packaging or packaging that can be recycled in the community, or avoiding single-serve disposable items when possible—consumers can reduce the garbage we generate and send manufacturers a message. In a number of cities, community service organizations, such as the League of Women Voters, sponsor environmental shopping tours to teach shoppers how to choose products that promote source reduction. (See *Resources* for information on environmental shopping tours.)

Exchange Programs. The adage "one person's waste is another's treasure" holds true even when the waste is unwanted paint, magazines, or furniture. Exchange programs give people a place to get rid of belongings they no longer want or to acquire items at little or no cost. The town of Wellesley, Massachusetts, provides waste exchange services at the town landfill and recycling center. Businesses also can participate in exchange programs. They can subscribe to product exchange catalogs or periodically check on-line computer inventories of products to exchange with other businesses in the community or region.

BAG A TREASURE FROM ANOTHER'S TRASH

Taking out the trash has been a little different in Wellesley, Massachusetts, for some time now. Not only do many Wellesley residents voluntarily drop off their trash and recyclables at the Wellesley Recycling and Disposal Facility, locally referred to as the "RDF," but they also combine disposal and recycling with rummaging for free used goods. They are able to do this because the RDF also

continued on next page

---

hosts a Waste Exchange Center consisting of a "take it or leave it" area for the exchange of reusable goods, such as old bicycles, lawnmowers, furniture, and other items that might be used by someone else, a book exchange designed for borrowing books or picking up books no longer wanted by others, and a Goodwill trailer. For those hard-to-find items, a note on the "swap shop information board" might just do the trick.

Consumer Activism. Consumers also can promote source reduction by writing letters, making phone calls, and circulating petitions to persuade state and federal officials to adopt source reduction policies or manufacturers to implement source reduction strategies. Talk or write to store managers to encourage them to add a bulk foods section and to carry products made from recycled materials and packaged with materials that are recyclable in your community. Serve on a local industry's citizen advisory board.

Source reduction is the first step. Figures 5 and 6 list additional activities we can do or encourage government and manufacturers to do to promote source reduction.

Figure 5. Government Source Reduction Checklist

Education

—Public outreach

—Distribution of source reduction data

—Awards to recognize community and industry source reduction efforts

—Seminars on source reduction techniques

Voluntary Programs

—Labeling on environmentally preferred products

—Waste audits, and product and packaging audits

—Life-cycle assessments

—Voluntary goals and standards for source reduction

Fees/Taxes

—Volume-based or weight-based disposal rates

—Landfill surcharges to provide revenues to finance source reduction projects

—Tax on products such as hard-to-dispose-of items (e.g., tires, motor oil), tax on certain disposable or environmentally harmful products

—Tax on materials such as virgin, all raw or certain toxic materials

—Deposits on certain products

Tax Credits/Subsidies

—Tax credits to businesses conducting source reduction activities such as redesigning or standardizing products, purchase of source reducing equipment, remanufacturing products, research and development

—Tax rebates to consumers and/or manufacturers of reusable, repairable, remanufacturable, or more durable products

—Government or industry grants to academia and nonprofits for source reduction research

—Loans or grants for pilot projects

—Subsidies for source reduction infrastructure

—Subsidies to businesses conducting source reduction activities

continued on next page

Administrative Actions

—Waste reduction officers to implement government programs

—Government research and development on source reduction techniques

—Procurement guidelines for source reduction activities by all levels of government

—Waste audits of government agencies

Regulatory Actions

—Elimination of restrictions or regulations that are obstacles to source reduction

—Elimination of virgin materials subsidies or policies that favor the use of virgin materials

—Waste audits

—Waste reduction planning requirements for large-quantity waste generators

—Disclosure of virgin material use, energy use, air and water emissions, and solid waste produced

—Disclosure of presence, amount or concentration of toxics

—Product warnings for toxics

—Product durability and reusability requirements

—Restriction on the use of certain toxic substances

—Ban on product, product materials or certain activities

—Ban on government purchase of certain items

Source: Adapted from World Wildlife Fund, *Getting at the Source: Strategies for Reducing Municipal Solid Waste.*

Figure 6. Manufacturers and Consumers Source Reduction Checklist

Manufacturers	Consumers
—Eliminate product/packaging or reduce amount	—Don't purchase product or reduce use
—Eliminate or reduce toxic substances in the product	—Purchase product with reduced toxics
—Switch to environmentally preferred materials or processes that use less energy and materials	—Purchase environmentally preferred product
—"Lightweight" or reduce products volume	—Purchase product with less packaging
—Produce concentrated product	—Purchase concentrated product
—Produce products in bulk or in larger sizes	—Purchase product in bulk or in larger sizes
—Combine functions of more than one product	—Buy multiple-use product
—Produce standardized models or styles	—Purchase fewer product models/don't replace for style
—Increase product life span	—Purchase more durable product
—Improve product repairability	—Maintain properly/repair instead of replace
—Produce for consumer reuse	—Purchase reusable product/ reuse product/donate to charity
—Produce more efficient product so less of a product is needed to provide the same service	—Purchase more efficient product or use product more efficiently
—Remanufacture product	—Purchase remanufactured product
	—Borrow, share or rent product

Source: Adapted from World Wildlife Fund, *Getting At the Source: Strategies for Reducing Municipal Solid Waste.*

Figure 7. In the early days, reuse and recycling were part of everyday life. (Courtesy of the National Solid Wastes Association).

RECYCLE AND COMPOST

No matter how successful source reduction is, there always will be trash to throw out. In the "old" days, reuse and composting were routine household activities. Before the 1920s, 70 percent of the nation's cities ran programs to recycle select materials. And during World War II industry recycled and reused about 25 percent of the waste stream.

Today, recycling and composting are again household activities for many people. Spurred by concern for the environment, the country's recycling and composting rate rose from 7 percent in 1960 to 17 percent in 1990, according to the Environmental Protection Agency.

RECYCLING

Recycling offers an alternative to the disposal methods of landfilling and incineration, which are discussed in Chapters 4 and 5. By reusing materials such as aluminum, paper, glass, and plastics, industry and government can achieve savings in production and energy costs, and can spare the environment from the negative impacts of the extraction and processing of virgin materials. Recycling also means there is less trash requiring disposal.

Despite renewed public support for recycling and pressure on local government to handle garbage more efficiently, recycling as a solid waste management option faces an uphill battle. While collecting and processing recyclable materials are becoming more popular, there is considerable debate over the economic feasibility of further recycling and the development of markets for recyclable products.

**RECYCLING
A THREE-STEP PROCESS**

Step One　　**Collect and Reprocess.** Materials are separated from the waste stream and prepared to become new raw materials.

Step Two　　**Manufacture.** Materials are used in the manufacturing of new products. (Generating fuel or power from burning garbage is not recycling.)

Step Three　　**Use.** The products made from recycled materials are purchased by consumers.

DOES RECYCLING MAKE ECONOMIC SENSE?

The debate on the economic viability of recycling continues. Many believe that recycling is the right thing to do and support legislative action to encourage recycling. Most states have passed recycled content laws, tax credits, disposal bans, or other types of legislation to encourage recycling. As a result of these efforts, many communities can produce figures to prove that recycling makes economic sense. The Institute of Local Self-Reliance recently published *Beyond 40 Percent,* a case study of 17 communities that are recycling and composting more than 40 percent of their garbage.

Advocates of recycling claim that since there is a steady supply of recyclables but only a sporadic demand, evaluating recycling solely on costs and on revenue from the sale of recyclables skews the picture. They

1889

Washington, D.C. Health Officer reports that, "Appropriate places for [refuse] are becoming scarcer year by year . . . "

argue that the economic equation must include the environmental, resource conservation, and energy benefits from recycling. They also maintain that the formula for evaluating recycling should include the "avoided" costs of disposal—the savings earned by *not* disposing of garbage in landfills or incinerators. Because the benefits of recycling are not currently captured in market transactions, advocates support legislative action to stimulate demand for recovered materials.

On the other side of the debate are those who support a free-market approach to recycling. If it is more expensive to recycle materials than to send them to a landfill or incinerator, they claim that recycling does not make sense. They argue that there are costs (dollars and resources) to recycling and trade-offs to make between using recovered versus virgin materials. They also assert that regulations designed to create markets for recyclables will not guarantee the efficient use of resources. Recycling makes sense in areas with high disposal costs and limited access to raw materials, but in other areas mandated recycling will only *increase* waste management costs and consumer product costs.

Both sides do agree on two issues: garbage fees should reflect the full costs of waste management, and consumers should be charged for what they throw out, either in volume- or weight-based collection rates.

Despite the recycling success stories and statistics showing the savings in resource extraction and production costs from recycling municipal solid waste, the feasibility of a local recycling program should be evaluated on a case-by-case basis.

Figure 8. Estimates of Energy Savings from Recycling
Using recyclables in the manufacturing process saves much of the
energy it would take to convert virgin materials for use and generates
less air pollution.

Material	*Energy Savings Through Use of Recyclables*	*Air Pollution Savings*
Aluminum	95%	95%
Plastics	88%	—
Newspaper	34%	73%
Corrugated	24%	—
Glass	5–30%	20%

Source: Adapted from *Garbage Solutions: A Public Officials Guide to Recycling and
Alternative Solid Waste Management Technologies,* as cited in *Energy Savings from Recycling,* January/February 1989; and Worldwatch Paper 76 *Mining Urban Wastes: The Potential for Recycling,* April 1987.

THE NEED FOR MARKETS

A maxim of the recycling business is "absent a market, a waste
material has merely been collected, not recycled." Markets for
recycled materials (also called secondary or recovered materials)
are the foundation for successful recycling programs. As the collapse of the market for old newspapers in the late 1980s showed,
successful recycling is contingent on the supply of materials *and*
the demand for their use. To ensure that materials collected will be
reused in production processes, there must exist an industrial
demand for recycled materials, as well as a consumer demand for
products made from recycled materials.

Many communities and states are assessing whether sufficient recycling markets exist or can be developed in their area.
Unfortunately, there is no easy answer. Market availability is often
a regional issue that is contingent upon the proximity to manufacturers that use recovered materials and to recycling plants. In the
northwest, communities have not had a problem recycling old
newspapers because of the large number of newsprint mills able to
use old newspapers and their shipping proximity to paper markets

in the Far East. Other regions, those not close to wood pulp sources, have experienced large market fluctuations for old newspapers. In the midwest, communities have little difficulty finding markets for green glass because of their proximity to green glass processors, but other parts of the United States have had problems find-

ing green glass container manufacturers to keep pace with the supply of imported beer and wine in green bottles. These and other regional examples demonstrate how markets are created. A market for recyclables depends on an identifiable source and supply of materials, a system to extract materials from the waste stream and deliver them to a recycling plant or manufacturer, a manufacturer that wants the materials and is able to remanufacture them into consumer products, and, finally, an existing or potential demand for the finished product.

Currently, industrial use of recycled materials is limited because of economic barriers that discourage recycling, including inadequate recycling capacity, low consumer demand, and certain federal policies that favor the use of virgin materials. For example, provisions in the federal tax code and federal energy subsidies that benefit primary material producers hamper the recycling industry. To overcome these disincentives and promote recycling, federal, state, and local governments can develop legislation that supports recycling and creates economic incentives to use more recovered materials, maintaining a balance between the demand and supply of recyclable materials.

Legislative Options

Supply . . . Legislative tools to increase the supply of quality recyclable materials include mandatory recycling collection laws, container deposit laws, and bans. Thirty-five states, the District of

WHAT'S REUSABLE AND RECYCLABLE IN THE WASTE STREAM?

Aluminum Cans Reprocess for can sheet and castings.

Animal Waste Use as fertilizer.

Automobiles Recover steel; reuse parts.

Construction Waste Reprocess for pressed cardboard, roads, and other construction projects.

Furnishings and Clothing Reuse by another person; reprocess for use by the textile industry.

Glass Refill or use cullet for jars, bottles, construction material.

Lead Acid Car Batteries Recover lead and plastic casings.

Motor Oil Re-refine as motor oil.

Other Metals Clean and reprocess as scrap and structural products.

Paper (mixed paper, high-grade paper, newspaper, cardboard) Reprocess for paper products such as newsprint, printing and writing paper, tissue, paper packaging, uncoated and coated paperboard, insulation.

Plastic Drink Bottles Reprocess for auto parts, fiberfill, strapping, new bottles, carpet, plastic wood, plastic grocery and trash bags.

Tires Retread for resale; reprocess as industrial and household products, gas, oil, and char.

White Goods (household and industrial appliances) Remove PCB-containing components and recover steel.

Wood Wastes (e.g., shipping pallets and boxes) Reprocess into wood chips for use as mulch or landfill cover.

Yard Trimmings Compost for landscaping.

Source: Adapted from EPA, *Recycling Works!*

Columbia, and Puerto Rico have adopted some form of mandatory recycling. Under this legislation a state or county sets recycling goals for targeted materials and timetables to achieve the goals (see Figure 10).

Container deposit/refund legislation or "*bottle bills*" can increase the supply of recyclables. Deposit legislation was first enacted to prevent litter-

1896

Waste reduction plants, which compress organic wastes to extract grease, oils and other by-products are introduced to the U.S. from Vienna. The plants are later closed due to their noxious emissions.

ing, but it has proven to be an effective method for collecting source-separated materials. (In most states with bottle bills, more than 80 percent of beverage containers are recycled.)[3] Deposit laws require retailers to pay a deposit for each container of soda, beer, or other beverage they purchase from distributors; consumers in turn pay the deposit to the retailers. When consumers redeem the cans or bottles at a store for the deposit, retailers can redeem the deposit from the distributor. Often retailers also receive a handling fee from the distributor. Nine states use this type of legislation to encourage recycling and reuse. Oregon was the first state to enact a bottle bill, in 1971, followed by Connecticut, Delaware, Iowa, Maine, Massachusetts, Michigan, New York, and Vermont.

California, apprehensive of the impact of a bottle bill on grocery stores, adopted a slightly different law, a beverage container redemption law. The law is a compromise between a bottle bill and an advance disposal fee. Consumers do not pay a deposit when they purchase a beverage but instead receive 2½– 5 cents, depending on the size of each bottle or can that they redeem at designated redemption centers. Critics argue that the bill discourages glass bottle reuse because broken bottles can be redeemed and that the redemption incentive is not high enough to encourage consumers to make a special trip to redemption centers. But the return rate was 82 percent in 1991, according to the state.

Although bottle bills only target 4 percent of the waste stream, they have boosted the markets for post-consumer glass, aluminum, and plastic, in particular polyethylene terephthalate (PET) plastic soft drink bottles. Despite the high collection rates bottle bill states have experienced, states considering similar legislation still encounter strong opposition from bottle and beverage industries because of the shift of collection costs from local government to industry and consumers.

Another method to increase the supply of recyclable material is a *disposal ban.* As discussed in Chapter 2, bans are controversial policy approaches. While disposal bans may be limited in their effectiveness to divert waste from disposal, they often do prompt product manufacturers to invest in solid waste management, especially in recycling. Disposal bans prohibiting the discarding of recyclable materials in landfills and incinerators are designed to encourage the recycling of certain materials or to keep toxics out of landfills and incinerators. Items frequently banned are lead-acid batteries, tires, yard trimmings, and used oil. By 1992 at least 34 states and the District of Columbia had instituted disposal bans of some kind. However, for this type of ban to work and illegal dumping to be prevented, recycling services must be provided and, most importantly, recycling markets must exist for the banned materials. For example, yard trimmings, which comprise about one-fifth by weight of the waste stream, are banned from landfills and incinerators by 16 states; municipalities in those states offer composting services as an alternative disposal method.

In New York City, the homeless are benefiting from the bottle bill. "We Can," a New York City nonprofit redemption center designed to serve homeless people, has paid more than $7 million in refunds and diverted more than 12,000 tons of glass, aluminum, and plastic from the waste stream since it opened in 1986. Every day about 500 people redeem cans and bottles at its two redemption centers. The organization also provides employment for 35 homeless or formerly homeless people.

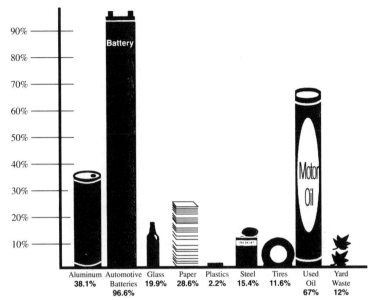

Figure 9. Recovery rates for Major MSW Components in 1990. Includes all materials in tires—rubber, textiles, ferrous metals. *Source:* EPA "Environmental Fact Sheet: Recycling Municipal Solid Waste, Facts & Figures"

. . . and Demand. Legislative options designed to encourage demand for recyclable materials include minimum recycled content mandates, utilization rates, recycled product procurement policies, and recycled product labeling laws.

Minimum recycled content mandates require that a certain percentage of a product be made from recovered materials, usually post-consumer materials (materials generated from residential and commercial waste versus pre-consumer materials, materials from industry processes that never reach the consumer such as broken glass from a manufacturing process). Proponents assert that minimum content rates guarantee markets for recyclable materials and

1898

▼

Colonel Waring, New York City Street Cleaning Commissioner, organizes the first rubbish-sorting plant for recycling in the U.S.

assure financing companies, which insure bonds for collection and processing facilities, that there will be a steady demand for recyclables. Opponents counter that content mandates artificially drive up prices for recycled products, impose burdensome reporting requirements on industry, and divert secondary materials from their most cost-efficient use to preassigned uses. They claim that industry needs flexibility to use materials in the most economical and technically appropriate manner.

However, mandatory recycled content legislation has proven to be successful in addressing the glut in the newsprint market caused by too many collected newspapers and not enough demand. A minimum content of recycled fiber in newspapers is required in 11 states (Arizona, California, Connecticut, Illinois, Maryland, Missouri, North Carolina, Oregon, Rhode Island, Texas, and Wisconsin) and the District of Columbia. Florida has taken a different legislative tack and placed a tax on virgin newsprint, giving newspaper publishers the option to use newsprint with 40-percent recycled content and receive a tax credit, or use paper with less than 40-percent content and pay a tax. Spurred by the move toward state adoption of mandatory newspaper recycled content laws, newspaper publishers in at least 12 other states (Indiana, Iowa, Louisiana, Maine, Massachusetts, Michigan, New Hampshire, New York, Ohio, Pennsylvania, Vermont, and Virginia) have reached voluntary agreements with state governments to use recycled newsprint. California also has adopted recycled content mandates for glass containers, plastic containers, trash bags, and fiberglass; Connecticut and Maryland have mandates for telephone books; Oregon has mandates for glass and plastic containers; Wisconsin has mandates for plastic containers; and Washington, DC has mandates for most paper and paper products.

Utilization rates are a more flexible regulatory option than minimum content standards because they offer a variety of ways for the targeted industries to comply. This approach requires manufacturers to use set amounts of recovered materials, but allows them to choose among several requirements; for example, use recovered materials in their own products or packaging, arrange for another manufacturer to use the recovered materials in its products or packaging (tradeable credits), reuse packaging for its original purpose, or reduce the volume or weight of a product's packaging. As with minimum content standards, utilization rates can be established by federal, state, or local governments through laws, regulations, or cooperative agreements.

1900

▼

Piggeries are developed in small- to medium-sized towns where swine are fed fresh or cooked garbage.

Procurement policies also can stimulate demand for recycled products. Purchases of goods and services by federal, state, and local governments combined represent approximately 20 percent of the U.S. Gross National Product (7–8 percent federal and 12–13 percent state and local). By purchasing durable, reusable, recycled, and recyclable products, government can use its purchasing power to support industries and businesses manufacturing and selling recycled materials. Typical procurement policies are "set-asides" that require the purchasing of a specified percentage of recycled products or "price preference" programs that allow government agencies to pay more for products containing recycled materials. In 1991, 16 states set aside a portion of their procurement budget to purchase recycled products, and 27 states allowed government offices to pay more for recycled products than comparable products made from virgin material. EPA requires all government agencies to buy recycled paper, re-refined oil, building insulation made with recycled material, and retreaded tires when using federal funds to purchase these products (if available at a reasonable price). Unfortunately, many of these procurement programs are still rather narrowly focused on purchasing small amounts of recycled paper rather than on large quantities of various recycled products. Many procurement officials report that their procurement programs are small because of the high prices of recycled products and because of problems with product availability and quality.

Another government approach to encouraging demand for recyclable materials is through *recycled-product labeling regulations.* If standardized product labels list the environmental attributes of the product, for example, the percentage of pre-consumer or post-consumer materials used in the product or packaging or availability of recycling facilities, then consumers can make educated buying decisions. (See Chapter 2.)

GERMANY STANDS BY THE PRINCIPLE "THE POLLUTER PAYS"

Germany has passed perhaps the most extensive packaging legislation of any country. In May 1991 Germany enacted legislation that requires the collection and recycling of 6–8 million tons of packaging annually. Packaging accounts for one-half the volume and one-third the weight of the country's household waste.

The German law is unique in that it requires those who create the packaging also to take it back. The legislation divides packaging into three categories:

One: Transportation packaging (pallets, corrugated containers, etc.) must be collected by producers and distributors, effective December 1, 1991.

Two: Secondary packaging—extra packaging to promote marketing or prevent theft (blister packs, tamperproof packaging, exterior cartons, etc.) must be taken back by retailers, if the consumer wishes, effective April 1, 1992.

continued on next page

Three: Primary packaging—sales packaging (liquid containers, boxes, etc.) must be collected by retailers or a privately funded collection company that guarantees minimum recycling rates for designated packaging materials, effective January 1, 1993. This is the most controversial mandate.

Afraid of becoming dumping sites for the country's discarded packaging, German retailers and 600 packagers and manufacturers of consumer products have created a packaging collection program. Duales System Deutschland (DSD), a private corporation, is establishing a nationwide system of drop-off facilities and curbside collection programs for packaging that carries a "Green Dot" logo. In order for a package to carry the Green Dot, the manufacturer must pay a licensing fee, which varies depending on the size and shape of the package; the average fee per container is 0.02 Deutsche marks. Packages with the Green Dot logo are then collected by DSD, and manufacturers are guaranteed a recycling market for their packaging. By March 1992 manufacturers had purchased 5,000 licenses allowing 40 billion packages to carry the Green Dot.

The program is funded from Green Dot licensing fees, which are expected to raise $1.2 billion per year in operating revenue. By 1995 DSD hopes to create 18,000 jobs. But as innovative as the Green Dot program is, it has raised certain concerns. One problem is that the program has generated a large over-supply of some recyclables, particularly plastics. Although industry has responded favorably to the program, recognizing that there is strong public commitment to recycling and that recycling is part of doing business in today's markets, the plastics industry believes that the recycling goals are too high. Currently, about 4 percent of the collected plastics is recycled; the

continued on next page

industry must achieve a 65 percent rate by 1995. Many plastic manufacturers think that incineration should count as a disposal option.

Many environmentalists believe that DSD licensing fees are not high enough to create an incentive to eliminate throwaway packaging and that the program does not solve the problem of "over-packaging" because it focuses on recycling rather than on source reduction. And some critics charge that the program's educational campaign suggests that all Green Dot products are environmentally "friendly."

Economic Options

An alternative to mandatory recycling is for government to provide economic incentives to industry and consumers to recycle voluntarily.

An option that is often discussed is a *virgin materials tax,* a federal tax levied on the use of virgin materials in the manufacturing of products and packaging. The objective of the tax is to conserve natural resources, reduce pollution, and encourage the use of recovered rather than virgin materials. The theory is that manufacturers would find it cheaper to use recovered materials than virgin materials and consequently increase the demand for recovered materials. However, this type of tax would be very controversial because of the economic impact it would have on the virgin materials industries.

A slightly different tack would be to provide *tax credits* to industries that use recovered materials. This option would attempt to counterbalance the tax credits received by raw material users. Under current federal law virgin material users get a tax credit called a resource depletion allowance and energy subsidies.

To offset these current federal policies, state and local governments already are offering tax exemptions to help recycling processors and manufacturers purchase equipment and other capital investments, rebates to companies that use recycled materials in new products, and grants and loans for new or existing recycling businesses.

PAY-AS-YOU-THROW

Want an incentive to recycle or a reason to consume less? Pay-as-you-throw may motivate individuals to pay closer attention to what they put in their garbage bins. Such was the outcome in Minnesota when mounting landfill tipping fees and increasing trash spurred the state to adopt a law requiring that trash fees increase as the volume or weight of the garbage collected increases. As a result, dozens of cities throughout Minnesota adopted pay-per-can collection systems. In January 1991 Duluth adopted this collection system. Residents are charged a monthly fee: $15 for one 20-gallon can; $20 for one 32-gallon can; $26 for two 32-gallon cans; and $33 for three 32-gallon cans. The city also has a special "mini can rate" of $10 a month for once-per-month collection of one 20-gallon can. In addition, and unlike most other cities, the trash fee includes $4 per month needed to cover the costs of curbside recycling.

The result in Duluth is less garbage at the curb on collection day, according to the president of the local trash haulers consortium. In the process, the new fee system has helped kick off curbside recycling and has rewarded those who put less garbage at the curb with lower rates. The revenues raised by the fees nearly pay for the entire waste collection system. However, some communities have experienced problems with similar programs. Some residents are not aware that they can subscribe to the cheaper service if they generate less trash, and some private haulers are concerned about a steady flow of revenue and frequent household service changes.

Consumer incentives to recycle include *variable can fees* combined with curbside collection of recyclable materials. A growing number of communities is turning to this type of collection system because it creates an economic incentive for residents to put less in their garbage cans and more in their recycling bins or bags. Variable can rates act as a signal to waste generators by requiring them to put their money where their garbage is. In Perkasie, Pennsylvania, recycling increased 150 percent in the first year of the city's pay-per-can/recycling program. (For further discussion, see Chapter 8.)

JOINING FORCES TO MARKET RECYCLABLES

Marketing cooperatives offer both rural and urban communities a new way to address fluctuations in recycling markets and other obstacles that prevent the growth of recycling programs. Local governments, nonprofit organizations, private recyclers, and businesses are joining forces to form marketing cooperatives to collect and sell recyclables. By working together members can better assure that the materials they collect and process will not be sent to a landfill or incinerator in the end.

The most common service provided by a cooperative is to identify markets and establish contracts or agreements with material buyers. Another service frequently offered is contracting with trucking firms for the transportation of recyclables to the end market. Some cooperatives provide education and training programs for members as a way to guarantee the quality of materials collected and to arrange for members to share processing equipment or storage. Cooperative marketing programs allow the co-op members to reduce recycling costs by pooling resources and sharing capital and operating expenses. And, the material buyers are more apt to sign a multi-year contract to buy recyclables from a cooperative because they are assured a steady supply of recyclables.

continued on next page

The main drawback to a cooperative marketing program is the loss of local control. As the co-op strives to supply uniformly marketable products, individual co-op members may lose flexibility in making marketing, equipment, and transportation choices.

Before a cooperative program is undertaken, the level of political and financial support from all potential participants, including elected officials, public and private recyclers, and community groups, should be carefully assessed. To cover the start-up costs of such a program, funding can be secured from state or federal government recycling and market development grants, membership dues, service fees for processing or marketing materials, in-kind services, trade groups, or foundations, and revenues from workshop fees and other training services.

Examples of successful cooperatives abound. The marketing cooperative in Searcy County, Arkansas, collects plastics for recycling. Ozark Recycling Enterprise is a nonprofit organization comprised of city, county, and nonprofit members. The co-op serves more than 85,000 people and has a one-year renewable contract with a plastics broker for high-density polyethylene (HDPE) and polyethylene terephthalate (PET) plastic packaging. Regional co-ops also are forming. In an effort to develop recycling markets in the southwest, cities in Arizona, Colorado, Nevada, New Mexico, and Utah formed the Southwest Public Recycling Association. This co-op serves 31 metropolitan areas. Its first contract supplies a tin ingot and steel manufacturer with bimetal cans from the region.

To encourage communities and states to launch marketing cooperatives, the New Hampshire Resource Recovery Association and the Wisconsin Department of Natural Resources provide information and technical assistance. The University of Wisconsin also maintains a database on cooperative marketing programs.

Figure 10. Source Reduction, Recycling, and Waste Reduction Goals in the United States. The goals listed are final goals; many states also have intermim goals.

State	Goals Source reduction	Recycling	Waste reduction	Local Plans Required
Alabama		25% by 1995		■
Arkansas		40% by 2000		■
California			25% by 1995* 50% by 2000 from 1990 baseline of total waste	■
Connecticut	12% by 2010	25% by 2010		■
Delaware		30% by 1994†		‡
Dist. of Columbia		45% by 1996		‡
Florida		30% by 1994		■
Georgia			25% by 1996 from 1992 per capita	■
Hawaii			50% by 2000	■
Illinois			25% by 1996 from 1991 baseline of total waste**	■
Indiana			50% by 2000 from 1991 baseline of total waste	■
Iowa			50% by 2000 from 1988 baseline of total waste	■
Kansas				■
Kentucky			25% by 2000 from 1993 baseline of per capita waste	
Louisiana		25% by 1992		■

continued on next page

Figure 10. *continued*

State	Goals			Local Plans Required
	Source reduction	**Recycling**	**Waste reduction**	
Maine	10% by 1994 from 1990 baseline of total waste	50% by 1994		■
Maryland		15–20% by 1994 20% by 1994 for state agencies		■
Massachusetts	10% by 2000 from 1990 baseline of total waste	46% by 2000		■
Michigan	5–12% by 2005 from 1989 baseline of total waste		50% by 2005	■
Minnesota		30% by 1996 for greater MN, 45% for Twin Cities		■
Mississippi		25% by 1996		■
Missouri		40% by 1998		■
Montana			25% by 1996 from baseline of 1991 total waste	
Nebraska			50% by 2002 from baseline of 1994 total waste	■
Nevada		25% by 1994		■
New Hampshire			40% by 2000 from 1990 baseline per capita	■

continued on next page

Figure 10.   *continued*

| State | Goals | | | Local Plans Required |
	Source reduction	Recycling	Waste reduction	
New Jersey	Cap waste generation of 1990 baseline by 1996, reduce by 2000	60% by 1995		■
New Mexico			50% by 2000 from baseline of 4 lb/day per capita	
New York	8–10% by 1997 from 1987 per capital baseline	40–42% by 1997		■
North Carolina			40% by 2001 from 1991 per capita baseline	■
North Dakota			40% by 2000 from baseline of 1991 total waste††	■
Ohio			25% by 1994 from 1989 baseline of total waste‡‡	■
Oklahoma				■
Oregon		50% by 2000***		■
Pennsylvania	No increase in generation from 1988 to 1997	25% by 1997		■
Rhode Island		15% by 1994 for residential	70%†††	‡
South Carolina		25% by 1997	30% by 1997 from 1993 baseline of total waste	■

continued on next page

Figure 10. *continued*

State	Goals			Local Plans Required
	Source reduction	*Recycling*	*Waste reduction*	
South Dakota			50% by 2001 from 1990 baseline of total waste	
Tennessee			25% by 1995 from 1989 base line per capita	■
Texas		40% by 1994		■
Utah				■
Vermont			40% by 2000 from 1987 baseline per capita	■
Virginia		25% by 1995		■
Washington			50% by 1995 from 1990 baseline of total waste	■
West Virginia		50% by 2010		■
Wisconsin				■
Puerto Rico		35% by 1995		

*Diversion from landfill, 10% allowed through incineration although the California Integrated Waste Management Board says any increase in incineration is unlikely.

†This is not a legislated goal, but a goal that the Solid Waste Authority, a quasi-government agency, hopes will be reached.

‡Solid waste planning is being done on a state (rather than a local) level for the District of Columbia, Delaware, and Rhode Island because they are so small.

**Counties with populations under 100,000 do not have to reach the 25% goal until 2000.

††Diversion from landfill, although the state has no solid waste incinerators and no proposals to build any.

‡‡Doesn't include yard waste, but does include incineration.

***This is a waste recovery goal, includes recycling, composting, and energy recovery (not waste incineration)

†††This is a waste processing goal, combines recycling and composting

Source: Adapted from INFORM, *Making Less Garbage: A Planning Guide for Communities,* Bette Fishbien and Caroline Gelb, 1993, pp. 26–29. Reprinted by permission.

THE BASICS OF RECYCLING

Recycling Collection Programs

Across the country four popular methods have emerged to collect recyclable materials: curbside collection, drop-off centers, buy-back centers, and deposit/refund programs. Curbside collection is the fastest growing recycling method in the country. Local programs exist in parts of every state except Alaska, Delaware (which is served by statewide drop-off sites), and Wyoming. In 1988 approximately 1,050 curbside programs collected recyclables; by the end of 1991 the number increased to 3,955 programs serving 71 million people.[4] Figure 11 outlines the pros and cons of these different recycling collection programs.

When recyclables are collected at the curbside they can be collected as mixed wastes, commingled recyclables, or source-separated recyclables. Collecting recyclables as part of the full municipal waste stream is referred to as *mixed wastes collection.* The advantage to this method is that it does not require changes in a community's current garbage collection system. Also, after collection, the community can select materials to recycle based on changing market conditions. A disadvantage is that the recyclables require extensive and costly sorting and cleaning—and some become too soiled to be cleaned.

Commingled recyclables describes the method of collecting all recyclables together, but separate from other trash. Collection programs vary from community to community. A community may provide a weekly or monthly pick-up service; truck operators may be responsible for sorting the materials into separate sections of the truck, or collecting the recyclables together to be sorted at a processing facility. This type of collection system has a lower contamination level than mixed wastes collection since some sorting has been done by residents or businesses. However, it does require an extensive public education program on what is recyclable and what is not.

Source separation describes the sorting of recyclables by residents and businesses before they are collected for recycling. Once collected, source-separated materials require the least complicated sorting equipment of all three collection methods. The advantage to this method is that the recycl-

1902

▼

79% of 161 U.S. cities surveyed in a Massachusetts Institute of Technology study provide regular collection of refuse.

ables are cleaner and therefore can be sold for a higher price. The disadvantage is that special collection vehicles are needed and collection takes longer because the truck drivers must sort the materials into separate bins on the truck. Both of these add costs to the program. A citizen education program also is needed.

After Collection

Once mixed recyclables are collected they are sent to a materials recovery facility (MRF) to be sorted and prepared into marketable commodities for remanufacturing. In 1992 there were 192 MRFs in operation and 19 more under construction. Although MRFs can be found in at least 34 states, they were mostly concentrated in the mid-Atlantic region in early 1992.[5] However, the number of MRFs built nationwide is growing at a substantial rate. These facilities primarily process residential waste, handling from 25–400 tons of garbage a day. To increase total revenues and use the full capacity of the plant (thus lowering operating costs), some MRFs have begun to process recyclables from commercial sources. Future MRFs will be designed to handle larger volumes and a greater variety of wastes, including construction and demolition debris.

Sorting of recyclables is done both manually and mechanically. Materials processed at MRFs include paper and non-paper commingled recyclables. Newspapers continue to be the major paper item processed at MRFs, although more MRFs are beginning to accept corrugated boxes, used telephone books, magazines, and mixed waste paper. The non-paper materials processed at MRFs

Figure 11. Recycling Collection Programs.

Option	Features	Advantages	Disadvantages
Drop-Off Center	Recyclables are self-hauled to drop-off center (e.g., landfill, transfer station) or drop boxes in parking lots Can be mobile Often sponsored by community organizations	Well-suited to communities where residents already haul their garbage to a disposal facility Easy to implement Low start-up and operating costs Accessible to commercial and industrial sectors	Inconvenient to public so participation rates can be low Less control over quality of materials Provides unsteady flow of materials Provides small reduction in waste stream Financial incentives are needed to increase participation (e.g., lower disposal fees, contests and lotteries at drop-off center) Subject to vandalism Requires citizen education
Buy-Back Center	Materials purchased from public/private sectors at redemption centers Cleaned and sorted recyclables are self-hauled to buy-back center Old newspapers, old corrugated cardboard, aluminum most common materials	Provides an economic incentive to participate Effective system for collection of aluminum Businesses and industry can sponsor buy-back centers	Inconvenient to public Good cash flow needed to purchase materials Has administrative and capital costs (labor, site, equipment) Provides small reduction in waste stream Requires citizen education

Curbside Collection	Materials collected at curb or alley Can collect commingled or separated recyclables Service provided by municipality or private firm Old newspapers, glass, aluminum, tin most common materials	Convenient to public Very effective method for residential recycling City has authority over collection system Collection can be integrated with garbage collection system Can provide high reduction in waste stream	Requires sophisticated collection system Time-consuming for garbage haulers High start-up and operating costs Requires citizen education
Bottle Bill/ Deposit Legislation	Consumer pays deposit on product and redeems deposit when product is returned	Increases market for post-consumer aluminum, glass, plastic High participation levels No costs to municipality	Affects small percentage of waste stream Imposes burden on retailers Strong industry opposition to new bottle bill legislation

Source: Adapted from EPA Region X, *Decisionmakers Guide to Recycling Plastics.*

Figure 12. Workers sorting recyclables at a materials recovery facility. (Photo by Hunter Gooch, Norcal Waste Systems, Inc.).

are typically aluminum and bimetal beverage containers (also referred to as used beverage containers or UBCs), tin-plated steel food cans, glass bottles and containers, and certain types of plastic bottles and containers. The plastics most often accepted are polyethylene terephthalate (PET) soda and water bottles and high-density polyethylene (HDPE) bottles, such as milk and laundry detergent bottles. The nonrecyclable items received at MRFs and materials that lose market value during processing, such as broken glass too small to separate by color, are typically combined and sent to a landfill or incinerator.

The processing of recyclable materials does have problems. For most materials, there is a significant difference between processing costs and sales revenues, for example. According to the National Solid Wastes Management Association (NSWMA), the average processing cost is approximately $50 per ton and the average value of recyclables is about $30 per ton. To compensate for this price difference, MRF operators can charge tipping fees (a set fee for each ton of recyclables or garbage dropped off at an MRF, landfill, or incinerator) or require processing contracts with communities or haulers.

MRF operators also must contend with potential negative environmental and health impacts. For example, MRFs generate increased truck traffic, air pollution, and noise levels. Also, plants that are poorly designed, operated, or managed can threaten the safety of workers. A recent Danish study showed that recycling workers are prone to skin, respiratory tract, eye, and gastrointestinal problems, with dust presenting the greatest risk to workers. To improve working conditions, some plants are installing proper ventilation systems and enclosing processing equipment. Other recycling processes, such as deinking newsprint and melting aluminum scrap, include pollution controls for air emissions and sludge. (See *Some Pollutants Generated During Recycling and Composting Processes* for a list of pollutants from processing recovered materials.)

How Do Materials Recovery Facilities Prepare Commodities for Market?

Materials recovery facilities (MRFs) prepare recyclables for markets according to the specifications established by user industries. Some samples follow:

Glass: To process glass into furnace-ready cullet so that it can be used directly in the manufacture of new glass containers:

- ◆ Only container glass is acceptable
- ◆ Glass must be separated by color into flint (clear), amber (brown), and green

continued on next page

> ◆ In flint glass, only 5 percent of the total load can be colors other than flint; in amber glass 10 percent; and in green glass up to 20 percent
>
> ◆ Glass must be free of any refractory materials; it will be rejected if there is more than a trivial amount of ceramic material
>
> ◆ Glass must be free of metallic fragments and objects, dirt, excessive amounts of paper, or large amounts of excessively decorated glass
>
> *Tin Coated Steel Cans:* To process cans acceptable to detinners so that tin and steel is able to be recovered:
>
> ◆ Not more than 5 percent aluminum contamination by weight
>
> ◆ Not greater than 10 percent paper labels remaining on cans or in load
>
> ◆ At least one can end removed
>
> ◆ Unwashed, but no food or liquid residue
>
> *Newspaper:* Shall consist of baled newspapers containing not less than 5 percent of other papers and total rejected material may not exceed 2 percent.
>
> **Source:** The United States Conference of Mayors, *Garbage Solutions: A Public Officials Guide to Recycling and Alternative Solid Waste Management*

COMPOSTING

Yard wastes constitute almost one-fifth of the municipal solid waste stream by weight; yard and food wastes combined account for about one-quarter of the total. Altogether, "nonrecyclable organics" (yard and food wastes, soiled, nonrecyclable paper, and other compostable materials) make up about 35 percent of the waste stream. Turning that refuse into a beneficial product would seem to be a good idea. Indeed, efforts to establish centralized composting facilities two decades ago promised "garbage into gold." But those facilities failed because they could not compete eco-

1904

▼

The first two major aluminum recycling plants open in Chicago and Cleveland.

nomically with landfilling. Today, while many see composting as an integral part of a comprehensive approach to municipal solid waste management, some observers emphasize that benefits accrue only when organics are source-separated. Otherwise, they caution, composting may be "garbage in—garbage out."

Composting is the biological process by which organic material, such as yard wastes, food wastes, and paper, is broken down by microorganisms (or microbes) in the presence of oxygen to form a humus-like substance—compost. Although organic materials decompose naturally, controlling key physical and chemical factors can accelerate the process and avert odors and other problems. Moisture, oxygen, available nutrients (especially the mix of carbon and nitrogen), temperature, and particle size all influence microbial activity and hence the composting rate.

Composting is a way to process that portion of the waste stream that does not lend itself well to other waste management approaches. Because nonrecyclable organic materials constitute so significant a portion of the waste stream, they take up considerable space if landfilled. In addition, the decomposition of organic wastes in the oxygen-starved environment of modern landfills produces methane, an explosive, heat-trapping greenhouse gas. In waste-to-energy incinerators, food and yard wastes tend to make inferior fuel because of their high moisture content.

The heat generated by microbial activity elevates the temperature inside a compost pile. The temperature needs to be maintained at a certain level (above 122°F) long enough to kill pathogens or weed seeds that might be present. Periodic turning ensures that the entire pile will be heated sufficiently and, at the

1911

▼

In Manhattan, Brooklyn, and the Bronx, citizens produce about 4.6 pounds of refuse each day. Yearly collections per capita include 141 pounds of wet garbage, 1,443 pounds of ash, and 88 pounds of dry rubbish.

same time, serves to dissipate heat from the interior, keeping temperatures from getting too hot (over 140°F) for the microbes needed for decomposition. Oxygen is critical to the process, as well. The worst odor problems are associated with anaerobic decomposition, which results from insufficient oxygen reaching the material within the compost pile. After an initial phase of rapid decomposition, the compost must be stored in piles or windrows (elongated piles) to stabilize or cure—that is, until decomposition completely ceases—before it is used.

High quality compost has a wide range of potential horticultural and agricultural applications. Compost can improve soil structure, enhancing the soil's capacity to hold moisture and nutrients, and reducing the need for chemical fertilizers. It can be used to reduce soil erosion and help restore degraded sites. Compost can be used by farmers, landscapers, home gardeners, land developers, and highway departments. Some compost producers report that they cannot keep up with the demand for their product. Producers of low grade, contaminated compost, on the other hand, often simply are stuck with it.

Composting can take place in individual backyards or in centralized facilities. While there are good reasons to pursue centralized composting, experience demonstrates that there are plenty of ways to do it wrong. Some facilities have been forced to close because of odors and other problems, for example, while others have had difficulty marketing the compost they produce. The challenge to waste management officials and citizens is to decide what is most appropriate to their area's needs and to design an effective program to produce a high quality product.

Choosing what to compost goes hand in hand with determining the appropriate composting technology. Composting systems range from simple piles or windrows to highly mechanized systems that send the refuse through shredders and enclosed rotating vessels called digesters, which accelerate the initial phase of decomposition. In between there are facilities with windrow-turning equipment and ones with forced air systems to assure adequate oxygen reaches the windrowed material. Most centralized facilities in the United States compost yard trimmings in windrows that are turned periodically as decomposition occurs. Because the highly mechanized "in vessel" digesters produce a finished compost more quickly (and may help control odors), they are attractive to high tonnage, mixed waste facilities. The cost of this technology is an important consideration for cities considering large-scale composting.

In some areas concerns have been raised about the health risks to workers and nearby residents posed by composting facilities. These concerns focus primarily on *Aspergillus fumigatus* (*A. fumigatus*), an airborne fungus that is commonly found outdoors in leaves, soil, and grass and indoors in basements, crawl spaces, and house dust. Although it is essentially ubiquitous, and poses no risk to healthy individuals, it can trigger a range of responses if inhaled by individuals with damaged immune systems. Consequently, *A. fumigatus* is classified as a secondary pathogen. Studies at sludge composting facilities indicate that workers have not suffered health problems that could be attributed to increased exposure. However, because levels of *A. fumigatus* at composting facilities are elevated above background levels, communities choosing to include composting as part of their waste management strategy must be prepared to effectively address community concerns.

PRODUCT STANDARDS AND MARKETS

Because compost is a product, waste management officials and planners must consider potential end markets in the planning stages as they determine the appropriate scale and degree of sophistica-

1914

▼

After a shaky start, incinerators catch on in North American cities. Approximately 300 plants operate in the U.S. and Canada.

tion of a composting operation. What standards do local or state regulations require for various end uses? What standards might buyers require?

At present there are no federal standards regulating compost made from mixed municipal solid waste or from yard trimmings. Of the few states that have established standards, some stipulate different minimum quality standards depending upon the end use of the compost. The state of New York, for example, requires that compost for residential use be of higher quality than compost used for land reclamation projects. Quality is determined chiefly by the level of chemical contaminants, especially heavy metals, found in the compost. Physical contaminants, such as plastic and glass, and other properties may have bearing as well.

Two distinct philosophies have emerged in the debate over compost standards. One advocates a risk-assessment approach wherein acceptable levels of toxic constituents such as heavy metals are established on the basis of their effect on animals, as determined by testing. The other advocates an ecological approach that establishes acceptable levels based on the amount of these constituents naturally present in the soil. With this approach, application of compost would not elevate the level of heavy metals or other potential contaminants above existing background levels.

Many who favor the risk-assessment approach think that the "no observed adverse effects level" (NOAEL) standards proposed by the federal government for sewage sludge compost should be applied to mixed waste compost as well. Others think these levels are far too high for compost that is to be widely used, or that sludge standards may not be appropriate for municipal solid waste compost. Further, they point out that risk-assessment tests are based in part on the soil's ability to bind the metals chemically, so the metals are not available for absorption by edible plants or by humans if the soil is consumed. (It is not uncommon for young children to eat soil.) However, it is believed that soil loses the ability to bind these constituents over a period of time. Although this period may be hundreds of years and may seem impossibly long to some Americans, it seems less impossible to many Europeans who may live in towns or cities that are hundreds of years old. Indeed, a number of European countries with

1916

U.S. produces 15,000 tons of paper a day, using 5,000 tons of old paper in the process, a 33% recycling rate.

Cities begin switching from horse-drawn to motorized refuse collection equipment.

a history of municipal solid waste composting are currently moving toward stricter standards and more source separation prior to composting, while some Americans favor the NOAEL standards. Source separation is seen by its advocates as the most effective means to keep contaminants out.

Because of the difficulty of testing for contaminants that are unlikely to be uniformly distributed, controlling inputs—along with testing the final compost—is necessary to ensure a safe, high quality compost. Because composting reduces volume, any contaminants present in the waste going in (called feedstock) and not broken down in the composting process will be concentrated in the finished product. Without mechanisms to screen toxics from going in, MSW compost may have "hot spots" (localized concentrations of pollutants) that do not show up in testing.

THE BASICS OF COMPOSTING

Yard Waste Composting

Yard waste programs constitute the large majority of composting operations in the United States. Many communities combine curbside leaf/yard waste collections or drop-off sites with public education campaigns to promote home composting. Some states have banned yard wastes from landfills, prompting communities to establish composting programs.

1920s

▼

Landfilling by reclaiming wetlands near cities with layers of garbage, ash, and dirt becomes a popular disposal method.

The advantages of backyard composting are obvious: residents set out less waste at the curb and take care of processing the portion they have diverted. Residents benefit by producing a useful and high quality garden product. In communities that charge per can for garbage collection, backyard composters are likely to save money as well.

Communities can encourage backyard composting by providing workshops and written instructions on the basics and benefits of composting, and some even subsidize the cost of compost bins. However, since not all households can be expected to participate, many communities have established centralized facilities to compost yard wastes.

Although yard waste composting poses relatively few risks, any centralized facility must be sited with the same general considerations given other municipal solid waste processing facilities, including the potential impacts of increased traffic, noise, and odors on the neighborhood, and the possibility of runoff from the site or generation of leachate (water that accumulates contaminants as it percolates through the compost; if uncontrolled, leachate can soak into the soil and eventually may threaten groundwater supplies).

Leaves are the simplest materials to compost. With a high carbon-to-nitrogen ratio and low moisture content, leaves decompose slowly and require only occasional turning. They tend not to carry significant pesticide residues or other toxic constituents and will decompose completely to produce a useful product. Grass clippings, on the other hand, have a low carbon-to-nitrogen ratio and high moisture content. They decompose quickly and can reek if not handled properly. Many programs urge residents to leave lawn clippings on the lawn rather than bagging them for curbside pickup. Many communities also mix leaves and grass clippings to produce a good compost feedstock. Brush from trees and shrubs must be chipped before it can be composted effectively; chipped brush can be used as a bulking agent for dense, wet material such as grass clippings. The relative amounts of leaves, grass, and brush will vary with the seasons, of course, and managers must anticipate and accommodate those changes.

Mixed Waste Composting

As landfills fill and incinerators generate growing community opposition, many local waste management officials are considering composting as a way to handle more than simply yard wastes. As of May 1992, 19 municipal solid waste (or "mixed waste") composting facilities were in operation in the United States, with 4 more under construction and 24 in advanced planning stages.[6] The bad news, however, is that several ambitious projects already have opened and closed in recent years. One, a $25 million facility operated by Agripost, Inc., in Dade County, Florida, which closed in early 1991 after only about a year in operation, had boldly attempted to compost the entire waste stream. Facility workers scanned the incoming garbage for hazardous wastes, pulled out the bulkiest items, and sent the rest through hammermills to be pulverized and then composted. Agripost sent only a tiny fraction of incoming wastes to the landfill as noncompostable residue, but ended up with a compost that many industry observers considered suspect. Although company officials state they were unable to keep up with demand for their product, the facility stockpiled large amounts of compost beyond the approved storage area in violation of its permits. The facility was shut down because of permit violations and persistent odor problems.

The key to an effective mixed waste composting program is devising an affordable and effective collection and separation system. Experience suggests that *source separation* is the key to retrieving compostable organics without wasting materials that can be recycled and without introducing contaminants that will jeopardize the final product.

Several communities with MSW composting facilities have started pilot projects to test various "wet/dry" separation strategies.

1942–45

▼

Americans collect rubber, paper, scrap metal, fats, and tin cans to help the war effort. The sudden surge of waste paper gluts market, and price drops from $9 to $3 per ton.

In the simplest wet/dry scheme, using two bins, all the wet garbage—the compostables—goes in one bin and everything else in the other. The task of sorting is straightforward, even with a few exceptions to the rules. (For example, soiled paper would also go into the wet bin.) The problem with this system is that if recyclables are to be recovered, they need to be pulled out at a transfer station or MRF, and they are likely to be of lower quality. In a three-stream wet/dry system, residents would sort wet organics into one bin, recyclables into another, and everything else into the third bin. The challenge with this approach is in achieving high and conscientious participation among residents. Either system requires a cost effective system to collect the sorted materials.

The advantages of collecting *unseparated wastes* are the savings in collection costs and the fact that more of the compostable waste stream is retrieved (along with everything else). These savings may be balanced, however, by the labor and capital costs to sort the wastes at the plant, the greater risk of toxic contamination if hazardous wastes are not detected and pulled out, and the likelihood that there will be fewer acceptable uses for the finished product. Generally, facilities that receive unseparated wastes attempt to pull out recyclables and hazardous wastes and compost the rest. At the end of the decomposition process, the compost is size-screened to remove larger pieces of plastics, glass, and other contaminants, which are sent to the landfill. The chances of introducing high levels of physical and chemical contaminants in this kind of system are significant and, as mentioned above, toxic "hot spots" in the compost may go undetected.

Another serious concern is that this approach squanders natural resources. Inevitably some recyclables will be missed, and those that are pulled out tend to be soiled and of lesser value. (Some facilities may skip pulling out recyclables altogether.) Paper that has been mixed in with the rest of the garbage is likely to be soiled, and therefore fit only for composting rather than recycling. Because recycling conserves natural resources used in the production of products, in addition to diverting materials from incinerators or landfills, it is preferred over composting for recyclable materials.

SOME POLLUTANTS GENERATED DURING RECYCLING AND COMPOSTING PROCESSES

As with virgin materials processes, most pollutants generated during recycling can be properly managed with pollution-control equipment.

Heavy Metals

Aluminum Recycling. As aluminum scrap is melted, heavy metals contained in painted labels, plastic, oil, and grease are burned off, creating air pollution. In the air, these metals form metallic chlorides and oxides, acid gases, and chlorine gas.

Steel and Iron Recycling. Steel and iron foundries produce solid wastes and sludge that contain lead, cadmium, and chromium. In some cases, these wastes are classified as hazardous due to their high concentration of heavy metals. Foundry furnaces also create air pollution. Electric furnaces emit lower levels of pollutants than coal furnaces, but both can emit high levels of particulate matter if the steel and iron scraps have high concentrations of alloys, dirt, and organic matter.

Paper Recycling. Wastewater and de-inking sludge from paper recycling can contain heavy metals such as lead and cadmium, which are used as pigments in printing inks.

Composting. Mixed waste compost tends to have higher concentrations of heavy metals than source-separated compost. Heavy metals found in compost include lead, chromium, copper, and zinc. MSW composted with sewage sludge also tends to have high metal concentra-

continued on next page

---

tions because the sewage treatment process removes heavy metals from the effluent and concentrates them in the sludge.

Dioxins

Paper Recycling. Wastewater and sludge generated at paper mills during the pulp bleaching process can contain dioxins.

Metal Recycling. Dioxins can be produced at secondary metals smelting plants. Dioxins have been detected in the air emissions of steel drum reclamation, scrap wire reclamation, and metals recovery from electronic scrap (e.g., telephones and circuit boards) plants.

Other Organic Chemicals

Paper Recycling. Inks removed during recycling often contain acrylics, resins, varnishes, and alcohols, some of which are discharged in the wastewater. Chlorine-based paper bleaching processes also can result in the discharge of various chemicals such as tetrachloride, methylene chloride, and trichloroethylene.

Plastics Recycling. Organic chemicals emitted from recycling plastics are a result of the solvent sometimes used to wash materials and remove contaminants. In some cases these solvents are toxic, such as 1.1.1-trichloroethane.

Composting. Grass clippings can contain organic chemicals from pesticides and nitrogen from fertilizers.

Chlorine and Sulfur

Chlorine and sulfur are chemicals found in many products, and chlorine is used in some recycling processes (e.g., paper recycling plants), so both chemicals are common residues found at recycling facilities. Common pollutants from these chemicals are hydrogen chloride and sulfur dioxide.

Source: Adapted from OTA, *Facing America's Trash: What's Next for Municipal Solid Waste?*

(See Figure 4 for health effects of certain pollutants.)

INCINERATION

To burn or not to burn? For many years, burning was the simplest and fastest way to reduce the amount of garbage. Until the late 1970s garbage and yard debris were typically burned at home and in private and municipal incinerators. With the passage of the Clean Air Act, backyard burning and other types of incineration were curtailed and new technology to burn our garbage was developed. Today, communities have the option of burning a large portion of their waste in modern incinerators with pollution control and energy production technology that was lacking in older facilities.

Incineration addresses our garbage dilemma by reducing the volume of garbage, but the process creates an ash that must be sent to landfills for disposal. Incinerators, also commonly referred to as waste-to-energy (WTE) facilities because most convert energy from burning garbage into steam or electricity, currently handle about 17 percent of the U.S. municipal waste stream. As demands for more landfill space increase, communities must assess the costs and benefits of incineration. By burning garbage, communities can reduce the volume of waste they send to landfills by 70–90 percent and can generate energy. However, communities must weigh these benefits against the costs of controlling and monitoring pollutants from air emissions, the disposal of incinerator ash, and the financing and siting of the facilities.

THE BASICS OF INCINERATION

Waste-to-energy incinerators are preferred over older incinerator models because they use an improved combustion process, have better pollution control technology, and produce energy from burning trash. In the United States in 1992, there were 145 WTE plants and 34 incinerators (plants that do not produce energy) burning municipal solid waste, plus more than 3,150 medical waste incinerators. By the end of 1992, 3 additional WTE plants were under construction, 37 were in the planning stage, and 9 were built, but not operating because of financial trouble or community opposition.[7]

Three standard models of incinerators are used in the United States, and each can be built as a WTE plant.

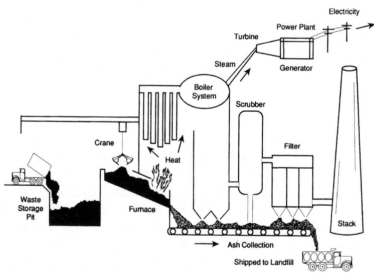

Figure 13. Mass burn waste-to-energy plant with pollution control system. *Source:* EPA, *Let's Reduce and Recycle.*

The *mass burn incinerator*, the most common model, receives garbage from which only the bulky or large noncombustible items, such as large logs and appliances, are removed; smaller items such as glass

1954

▼

Olympia, WA, enacts the first pay-per-can program.

bottles, batteries, and metal cans are left in the mixture to burn. Mass burn facilities have the capacity to burn 100–3,000 tons of garbage per day.

At a typical mass burn incinerator, mixed garbage is emptied on the tipping floor by incoming trucks and then moved into a large pit, where it is stored until burned. An overhead crane is used to mix the garbage in the pit to provide a uniform fuel mixture and then to load it into hoppers leading to the furnace. Fans in the furnace floor and walls provide air for combustion. As the garbage burns, the ash on the bottom of the furnace is removed through a water-quenched conveyor system and emptied into a storage area or transport trailer. Some plants remove and recycle the larger pieces of ferrous and nonferrous metals left in the ash. The recycling of other materials usually occurs before the garbage is collected.

A *modular incinerator* is similar to a mass burn facility, but typically has a smaller burning capacity. This type of incinerator can handle 15–400 tons of trash a day. It is often built in units at a factory and then transported to the facility site.

A *refuse-derived fuel* (RDF) *incinerator* burns garbage that has been preprocessed and sorted at the incinerator or at a processing facility. Recyclable materials such as ferrous metals, aluminum, and glass are removed and processed for sale or disposal, and the remaining garbage is shredded or processed into a uniform fuel, such as pellets, to be used in specially designed boilers or to be mixed with coal. This fuel has an energy value comparable to high grade coal. By October 1992 RDF facilities accounted for 20 percent of the WTE plants in the United States. But many incinerator operators are reluctant to use this type of fuel because the

Figure 14. Operating U.S. Municipal Waste Incinerators as of October 1992

Type of Technology	Number of Facilities	Average Daily Design Capacity (Tons Per Day)
Incinerator	34	6,957
Mass Burn Waste-to-Energy	65	65,534
Modular Waste-to-Energy	48	5,283
Refuse-Derived-Fuel (RDF)	16	25,310
RDF Combustion	13	5,150
(Burns RDF plus other fuels)		
TOTAL	176	108,234
Fluidized-Bed-Incinerator	3	—
(Burns RDF plus other fuels)		—
Medical Waste Incinerator*	3,150	—
Onsite	3,000	—
Offsite	150	—
*This figure is from 1990.		

Sources: Integrated Waste Services Association, Environmental Protection Agency.

boilers must be altered to burn it and the fuel can smell when stored for a long period of time.

A WTE technology new to the United States is *fluidized-bed incineration.* Data show that this technology has higher combustion efficiency and lower emissions of nitrogen oxides, sulfur dioxide, and dioxins than conventional incinerators. A fluidized-bed incinerator injects refuse-derived fuel into a loose moving bed of limestone and sand suspended above the furnace floor by an upward flow of air. The sand and limestone help to distribute heat evenly throughout the furnace, thereby limiting the formation of pollutants caused by incomplete combustion such as dioxins. In addition, the limestone tends to neutralize acids, thus lowering the emission of acid gases such as sulfur dioxide. However, the limestone process creates significantly more ash than other incinerators. Only 3 U.S. incinerators, in Minnesota, Wisconsin, and Washington, use such a system to burn a combination of garbage and coal or woodwaste. The first fluidized-bed incinerator designed to burn only garbage is under construction and scheduled to open in Cook County, Illinois, by 1995. Although 40 percent of Japan's incinerators use this system, and Norway and Sweden each have a facility, fluidized-bed incineration has not been widely promoted in the U.S. and is often more expensive than other incinerator technology.[8]

ENVIRONMENTAL AND HEALTH CONCERNS AND TECHNICAL SOLUTIONS

The decision to build an incinerator often incites considerable community concern and debate over the possible environmental and health risks. Typically, debate focuses on the issues of air pollution and ash.

Air Emissions

Incineration unavoidably creates air pollution. As with any other burning process, the smoke or emissions must be controlled to minimize environmental and public health risks. Concern over health risks posed by air emissions from incinerators—in particular dioxins, lead, cadmium, and mercury—has prompted a dramatic improvement in air pollution technologies over the past 10 years.

Heavy metals and dioxins are of concern because they can have adverse health effects on humans as well as on plant and animal life. For example, cadmium and dioxins are potential carcinogens, and mercury can damage the central nervous system and kidneys. Air pollutants generated by incinerators also can include particulate matter, carbon monoxide, nitrogen oxide, and acid gases such as sulfur dioxide and hydrogen chloride. Materials that contribute to these harmful emissions include household batteries that contain cadmium and mercury, lead-acid vehicle batteries, and certain plastics such as polyvinyl chloride (PVC) plastics that contain lead and cadmium. However, technical work is under way to reduce and phase out the use of mercury in household batteries and the use of lead and cadmium in plastics. Yard wastes and other materials with a high moisture content that cause the furnace to burn inconsistently also can produce harmful emissions.

Although modern incinerators use sophisticated air pollution control technology, emissions must be carefully monitored and controlled. EPA estimates that more than 95–99 percent of particulate and organic pollutants can be removed from air emissions if certain pollution prevention steps are followed. To limit the formation of dioxins and furans, an incinerator must be operated properly and maintain a constant burning temperature. To control air emissions and fly ash, an incinerator should use scrubbers followed by a baghouse (also called a fabric filter) or an electrostatic precipitator (ESP).

THE TROUBLE WITH MERCURY

Mercury tends to elude incinerator pollution control technology. When products such as household batteries, fluorescent lights, paints, or electronics are burned, mercury is released as a gaseous vapor. To capture vaporized heavy metals with pollution control equipment, the metals must condense onto ash in the flue gas. The difficulty with mercury is that it condenses at a much lower temperature (300–400°F) than other heavy metals. If the flue gas is not properly cooled, mercury will remain a gas and pass through scrubber and baghouse systems (electrostatic precipitators remove little if any mercury) and be released into the air.

Solutions for reducing mercury emissions include removing products containing mercury before incineration, reducing the mercury content in products, and installing activated carbon pollution control equipment that can capture more than 50–70 percent of mercury released during the burning process.

Incinerators can use either a dry or a wet scrubber system. Scrubbers are important because they control acid gases and cool the flue gas, causing pollutants to condense before entering the baghouse or ESP. Dry scrubbers or dry sorbent injection (DSI) scrubbers immerse the emissions in a very fine powder sorbent to condense the acid gases, allowing them to be collected in a baghouse system or ESP. The advantages to the DSI scrubber system are that it has low capital costs and can be installed easily in existing facilities. Wet scrubbers use the same concept as dry scrubbers, but spray a lime and water mixture into the scrubber to convert the gases into liquids and solids for collection in the electrostatic precipitator. Wet scrubbers generally are not compatible with a baghouse system.

A baghouse is composed of a series of porous bags through which flue gases, but not particulates, can pass. It functions much like a vacuum cleaner bag. Baghouses are replacing electrostatic precipitators as the pollution control technology of choice because they collect more particulate matter and are better at trapping very fine particles (heavy metals and organics tend to attach to smaller particles). An ESP removes particulate matter by negatively charging the particulates as they pass through the unit and collecting them as they exit the unit on positively charged plates.

Facility owners can install the pollution control technology of their choice as long as it meets EPA emission limits. EPA helps plant developers and managers to select the most appropriate technology by identifying the best available control technology (BACT) for new plants, or for retrofitting old plants.

One consequence of improving air emissions is producing a more toxic ash, because heavy metals are transferred from the air to the ash. The ash then must be properly treated and disposed of safely.

Ash

Ash is a substantial by-product of incinerated garbage. An incinerator that burns 1,000 tons of trash per day can generate between 200 and 250 tons of ash a day as residue.

1959

American Society of Civil Engineers publishes the standard guide to sanitary landfilling. To guard against rodents and odors, the guide suggests compacting refuse and covering it with a layer of soil each day.

Under current federal law, municipal solid waste incinerator ash is not regulated as a hazardous waste (waste that exceeds ignitable, corrosive, reactive, and toxic regulatory limits). Congress has allowed waste-to-energy facilities to be exempt from RCRA's hazardous waste regulations under Subtitle C if they burn municipal solid waste (household and nonhazardous commercial and industrial wastes). It is unclear whether Congress meant to exclude these facilities if the ash they generate tests hazardous. EPA's recommendations on whether to treat municipal waste incinerator ash as hazardous have been inconsistent. A September 1992 EPA memorandum stated that municipal waste incinerator ash should be exempt from hazardous waste regulation under Subtitle C and regulated as municipal solid waste under Subtitle D, reversing the agency position taken in 1985. However, EPA's position is only a recommendation and therefore the issue is still open to litigation should a municipal solid waste incinerator generate hazardous ash.

Incinerators generate two kinds of ash: fly ash and bottom ash. Fly ash includes charred paper, cinders, soot, and other materials that rise with the hot gases as garbage is burned and are often captured by air pollution control equipment in the stacks. Public concern over fly ash focuses on the heavy metals and dioxins that tend to attach to ash particles. Fly ash accounts for only 10–25 percent (by weight) of the total ash generated by an incinerator, but it is often more toxic than bottom ash.

Bottom ash, the material that is left in the combustion chamber after burning, is composed of noncombustible and incompletely burned garbage. Bottom ash accounts for 75–90 percent (by weight) of the ash generated by an incinerator. Most facilities combine the fly and bottom ashes and then quench the ash in water to cool it and prevent the fine particles from blowing away when the ash is handled and transported. The composition and toxicity of both types of ash depend on the contents of the waste burned and the efficiency of combustion. The combustion efficiency of an incinerator is determined by the temperature (high combustion temperature), time (long enough time for

1961

The Garden State Mill in Garfield, NJ, is the first mill in the U.S. to deink old newspaper to make newsprint.

waste to be exposed to high temperatures), and turbulence (sufficient mixing of the waste with oxygen).

Incinerator ash can be potentially harmful to humans and the environment if improperly managed. Municipal solid waste ash inherently contains toxic elements because it is composed of the remains of household, business, commercial, and sometimes construction and medical wastes, some of which are toxic. It can contain high levels of heavy metals and dioxins. When metals are burned they vaporize and attach to ash particles. Heavy metals often found in ash, especially in fly ash, include cadmium, lead, mercury, arsenic, beryllium, zinc, and copper. Dioxins, which are formed during the incineration process, can be found in trace concentrations in fly ash. Incinerator operators can minimize the formation and release of dioxins by using high, steady furnace temperatures (at least 1800°F) and by cooling the smoke.

Another reason ash must be properly managed is because it is lightweight and small and therefore can be dispersed easily in the surrounding environment where it can be ingested and inhaled. In general, burning increases the "bioavailability" of substances. For example, an incinerator quickly transforms a car battery into ash and pieces of metal, making the toxic components of the battery more accessible to humans and the environment if the ash is not properly managed. Ash also can leach heavy metals and threaten surface and groundwater sources.

Proper management practices include covering ash in transport trucks and storage piles. Mixing lime with ash and water to form a substance similar to concrete also can help to control the ash.

Environmental organizations have developed recommendations for sound ash management. Four basic steps suggested by the Environmental Defense Fund, for example, include: (1) manage fly ash and bottom ash separately, since fly ash is generally more toxic; (2) dispose of all ash in secure monofills, i.e., landfills that are designed specifically for only one waste, in this case incinerator ash; (3) chemically or physically treat ash to reduce its toxicity before disposal; and (4) keep materials containing heavy metals out of incinerators.

FROM ASHES TO . . . BUILDINGS, ROADS, AND REEFS

Nationwide, burning garbage produces an average of 8 million tons of ash per year. The abundance of ash, combined with its high costs of disposal, has led several communities and private companies to explore the possibility of using ash instead of sending it off to landfills.

A variety of pilot projects are under way to test the use of incinerator ash to make ash blocks for construction (similar to cinder blocks) and for use as road construction material and landfill covers. One of the more innovative pilot projects is the use of an ash and cement mixture to build artificial reefs off Long Island, New York, and Florida. Despite the headway being made in ash research and development, little action has been taken to commercialize its use. However, EPA's recent exemption of MSW incinerator ash from hazardous waste regulations under RCRA could change this.

Proponents and opponents of using ash can agree on two facts: heavy metals are common constituents of ash (particularly lead and cadmium), and there is no method that can eliminate all heavy metals from ash. As a result, ash utilization projects have focused on ways to permanently

continued on next page

lock these metals in the ash. This is where the controversy arises. Should a product be used commercially if it is known to contain heavy metals? There are areas of common ground among many technicians, academicians, and industry representatives who support using incinerator ash:

- ◆ Only bottom ash should be used because it has a lower concentration of heavy metals than fly ash.
- ◆ New tests designed specifically for ash utilization are needed. Tests should evaluate ash toxicity, leaching, and product degradation over time.
- ◆ Further research and testing of the different uses of ash are required.

Areas of concern include the fact that it is impossible to provide long term control over ash when it is used in the open environment. For example, ash blocks used in the construction of a building may not pose a risk while the building is standing, but what happens when the building is demolished? During demolition, dust containing heavy metals could pose health and environmental risks. By using ash are we simply delaying the risk it poses instead of eliminating it by disposing of it in safe and contained areas?

Traffic

As with other disposal facilities, incinerators generate truck traffic. At large incinerators, several hundred trucks a day pass through the gates to drop off garbage and haul ash to landfills. Communities considering incinerators should examine the potential impacts from noise and air pollution caused by truck traffic. Unlike landfills, which tend to be located in remote areas, incinerators, which are comparable in size to a power plant, often are sited close to communities to facilitate the transport of garbage.

ECONOMIC CONSIDERATIONS

Incineration is capital-intensive and requires substantial financial commitment from communities that decide to include this technology in their integrated waste management plan. The cost of building an incinerator is comparable to that of an airport or stadium. To help offset some of these costs, many communities are choosing to build waste-to-energy plants to produce steam or electricity for sale; however, these revenues will not recover the full costs of building and operating the incinerator. Factors that affect the total cost of an incinerator include facility size (the tons per day capacity), technology, location (which affects labor, transportation, construction, and land prices), pollution control technology, type of financing, ownership, and cost of ash disposal. Most incinerators are financed by a combination of bond issues, tax incentives, and revenues from the sale of steam or electricity. (For more detailed information on financing municipal solid waste facilities, see Chapter 9.)

POLICY RESPONSES TO INCINERATION

Incinerators are subject to federal, state, and local regulations. The principal federal statutes controlling municipal solid waste incinerators are the Clean Air Act, the Resource Conservation and Recovery Act (RCRA), and the Public Utilities Regulatory Policy Act (PURPA). Of these statutes, the Clean Air Act is the most comprehensive. In 1990 the Clean Air Act was amended to require EPA to develop more stringent and specific regulations for air emissions. The amendments require many industries, including incinerator operators, to retrofit existing facilities to meet the new standards. As a result of the amendments, in February 1991 EPA issued New Source Performance Standards and Emission Guidelines to cover both new and existing incinerators with capacities greater than 250 tons per day. These guidelines set limits for dioxins and furans, carbon monoxide, sulfur dioxide, nitrogen oxides and particulate matter, and heavy metals (mercury is not included in the guidelines). The guidelines also require the installation of the best available control technology (BACT) which includes scrubbers and electrostatic pre-

1965

▼

The first federal solid waste management law, the Solid Waste Disposal Act, authorizes research and provides for state grants.

cipitators or fabric filters (baghouses). It is the responsibility of each state to ensure that these standards are met and enforced. EPA estimates that the guidelines will reduce nationwide emissions of these pollutants, other than mercury, by 73–99 percent. Additional requirements called for in the 1990 Amendments to the Clean Air Act may extend incinerator regulations to include incinerators with capacities smaller than 250 tons per day.

Although RCRA is the primary federal statute that addresses municipal solid waste, specific regulations for municipal waste incinerators do not currently fall under RCRA's domain. Only hazardous waste incinerators and hazardous incinerator ash are regulated under RCRA. The absence of incinerator regulations means that municipal solid waste ash is regulated under RCRA's Subtitle D landfill criteria, unless the ash is classified as hazardous, and air emmissions are regulated under the Clean Air Act. EPA has developed general guidelines for the operation of MSW incinerators, and encourages regulation of municipal incinerators and ash within state solid waste plans under Subtitle D.

PURPA was passed in 1978 to provide economic incentives for the development of renewable resources for electrical power production. Under this legislation, local utilities must purchase power generated by WTE incinerators at a rate equal to the utility's avoided cost (the cost to produce the power itself or purchase the same power from another source).

Incineration also is controlled under state and local government regulations and permits. States, based on criteria set by the federal government, enforce air pollution emissions regulations. Some states also regulate the disposal of ash. Local governments can control incinerators with zoning ordinances and building codes.

DO INCINERATION AND RECYCLING COMPETE FOR RESOURCES?

Yes. Many proponents of recycling argue that incineration does compete with recycling programs for trash as well as financial and political support from local government.

There are facts to support this argument. Approximately two-thirds of the garbage burned in incinerators is recyclable or noncombustible. In order for a mass burn incinerator to run efficiently and to be economically viable, it needs a steady stream of garbage to meet its "tons per day" requirement (ranging from 250–3,000 tons per day, depending on its size). Many communities have had to enter "put-or-pay" contractual agreements with incinerator companies to guarantee that a minimum level of waste will be delivered to an incinerator; if the community falls short of its contracted level it must still pay a fee. As a result, the recyclable materials in the waste stream often are automatically sent to the incinerator regardless of the availability of recycling markets.

Incineration is a business, but the community does not always share in its profits. Once an incinerator is built, tipping fees, revenue from the sale of energy and recovered materials, and disposal contracts supplemented by tax breaks and government subsidies assure a steady flow of revenue to the operator. But because of high capital costs to build an incinerator and the possibility of lost revenue through tax breaks, a community may be left with little money in the city coffers to support its recycling program.

continued on next page

Recycling is a public service and a business. Once a recycling program is implemented (often at the demand of area residents), municipal funding and sales of recovered materials are its primary revenue sources. The day-to-day operating costs of an incinerator may be lower than those of a recycling program, but these savings are outweighed by the very high capital costs of incineration, and the energy savings from recycling. A recycled ton of aluminum saves energy equivalent to about 37 barrels of oil; a recycled ton of newspaper saves energy equivalent to 4 barrels of oil. Burning the same amount of aluminum *costs* about one-fifth of a barrel of oil; burning a ton of newspapers generates energy equivalent to two and one-quarter barrels.

No. Proponents of incineration cite other facts indicating that incineration and recycling programs can be compatible. Of the 145 waste-to-energy plants in operation, 60percent are located in communities that have active recycling programs. Proponents claim that both incineration and recycling can work together to reduce the volume of garbage ultimately sent to landfills. The problem of limited landfill capacity cannot be solved by recycling only.

Recycling and incineration can in fact be good partners. By working in tandem, recycling and incineration can process different segments of the waste stream. Recycling and composting programs can collect materials that burn poorly and generate little energy, such as aluminum, glass, and yard wastes. Incinerators can burn materials that have high fuel content, such as mixed paper and plastic. Incinerators also can support recycling efforts by subsidizing program costs through tipping fees.

Due to limited markets and infrastructure for processing recyclables, most recycling programs handle less than 50 percent of a community's waste. The trash that is not recycled can be put to work generating energy for the community to use.

MEDICAL WASTE INCINERATION

The summer of 1988 was ushered in with beach closings and pictures of medical waste washing ashore along the east coast of the United States. The resulting outcry prompted federal and state governments to take steps to track and regulate the management of medical waste.

Each year about 500,000 tons of medical waste are produced in the United States, mostly by hospitals. Medical waste includes pathological wastes and blood from surgery and autopsy, sharps (e.g., syringes, scalpel blades, and broken glass), and wastes from laboratories and research facilities such as contaminated lab clothing, equipment, and animal carcasses. Although medical waste accounts for only 0.3 to 2 percent (by weight) of the total municipal solid waste stream, it poses serious potential environmental and health risks if handled improperly.

Infectious, or "red-bag" waste, represents approximately 15 percent of the medical waste produced, and must be sterilized or decontaminated by the medical facility before it can be disposed of. Once the red-bag waste is treated, it can be handled like other municipal waste.

It is estimated that 70 percent of all hospitals run their own on-site incinerators, typically burning red-bag and all other medical waste. For example, hospitals also incinerate many types of chemical wastes. On-site incineration is a popular medical waste disposal option because it treats practically all types of waste generated by hospitals and avoids the problem of disposal facilities rejecting the waste.

Environmental Impacts

Hospital incinerators raise many of the same environmental concerns as municipal solid waste incinerators. Moreover, medical incinerators are often located in residential areas, and many old incinerators in operation are exempt from national emission control standards and best available control technology requirements at is time. Many states have taken the initiative to regulate medical waste incinerators because most medical incinerators do not fall within the current new federal source performance standards. Not until 1995 will medical incinerators with a capacity less than 250 tons per day be included in these standards.

The primary environmental concern arising from medical incinerators is toxic emissions of heavy metals and other chemicals. Pollutants frequently emitted from medical waste incinerators include cadmium, hydrochloric acid, and dioxins. Cadmium is used in the dyes for red-bags and in certain medical supplies such as plastics and batteries. Hydrochloric acid can be produced by burning products containing chlorine such as paper and PVC plastic. Of the few statewide studies conducted on air emissions from medical waste incineration, a recent California Air Resources Board (CARB) study is one of the most extensive. After testing 8 facilities, the CARB concluded that dioxin emissions from medical waste incinerators have the potential to pose the greatest risk of all identified sources of dioxin. The emission tests also revealed that all 8 facilities were emitting cadmium and hydrochloric acid at levels high enough to pose a potential risk. As with municipal solid waste incinerators, air emissions from medical incinerator stacks can be greatly reduced by retrofitting with proper pollution control equipment, an expensive undertaking.

1968

President Johnson commissions the "National Survey of Community Solid Waste Practices," which provides the first comprehensive data on solid waste since the 1800's.

Reynolds begins "buying-back," paying people for used aluminum— 8 cents per pound.

Alternatives to Medical Incineration

Less expensive treatment methods do exist, but few provide as many advantages as incineration. Steam sterilization (autoclaving)

1969

▼

Coca-Cola initiates the first life-cycle analysis study in the U.S. by examining the energy consumption of alternative beverage containers.

for medical waste is a commonly used alternative method. The process of steam saturating and high temperature heating sterilizes infectious waste so it can be disposed of in landfills, ground up and placed in sewers, or incinerated. Other treatment methods that disinfect and sterilize, but do not reduce the volume of waste include microwaving, thermal treatment, and chemical disinfection.

Economic Considerations

In an attempt to comply with state air quality regulations, hospitals and other medical waste producers are comparing the costs to retrofit their incinerators with the costs of using alternative treatment and disposal methods. Many are finding that it is cheaper to send their waste to large regional medical incineration plants despite the facts that they lose control over the disposal process and that their accident liability increases when the waste is transported.

Policy Responses to Medical Waste

Due to public concern over the mismanagement of medical waste, the federal government has conducted studies on several aspects of its handling and disposal. But until EPA's updated new source performance standards for medical waste incinerators are passed, the implementation of effective medical waste incinerator regulations will remain in the hands of state and local governments.

As of 1991, 25 states had passed regulations or permit guidelines specific to medical waste incineration, while 13 other states were proposing regulations and permit guidelines. In addition, 9 states have placed moratoriums on the construction of new medical incinerators and encouraged the use of other treatment methods.

The types of pollutants and allowable emissions greatly vary from state to state. One area of concern is the flexibility states have in selecting the minimum size of incinerators to regulate. In some cases, smaller, but not necessarily safer, incinerators fall outside state regulation.

FIVE

LANDFILL

Since the beginning of human history people have been dumping their garbage, and landfills will continue to be a component of waste management plans for the foreseeable future. Even after source reduction, recycling, composting, and incineration, there will still be garbage to dispose of.

For centuries we gave little thought to how our trash might affect the environment. Now concern about groundwater pollution, surface water contamination, methane gas generation, and other negative impacts of landfilling waste have led to mounting public opposition to "local dumps" and prompted the transformation of dumps into modern sanitary landfills. Instead of communities hauling their garbage to the local ravine, sand pile, or old quarry at low or no cost, they are hauling it to high-priced modern landfills designed to protect the environment and the public. Sometimes garbage is shipped hundreds of miles to avoid high local landfilling fees or local opposition to a new facility.

Over the past 20 years the number of landfills in the United States has decreased significantly. Yet, we continue to produce an ever-increasing amount of garbage. In 1979, nearly 20,000 landfills were accepting municipal solid waste; by 1991, this number dropped to about 5,800. Currently, it is estimated that there are just over 5,300 landfills in operation, and approximately 70 percent of the garbage we generate winds up there.[9]

The sharp decline is the result of landfills either reaching their capacity, or closing because they do not meet state or federal landfill design and operation standards. The number of landfills is expected to continue to decline as a result of stricter federal landfill standards under the Resource Conservation and Recovery Act (RCRA). Many of the smaller local landfills are being replaced by larger local and regional landfills, so the country's overall landfill capacity is not declining. However, the number of landfills closing does raise questions. Can the new larger landfills meet the needs of communities across the country? And will the larger local and regional landfills be in the right places? It may be easier to site and build a landfill in the open spaces of Texas than in the crowded northeast, but that does not help communities in the northeast faced with decreasing landfill capacity unless they are willing to pay costly disposal fees for transporting the garbage thousands of miles. And it does not address the fact that people who live in those open spaces do not want to be the dumping grounds for other people's garbage.

The siting of new landfills is hampered by the poor environmental track record of older dumps. More than 20 percent of the 1,200 cleanup sites on the Superfund National Priority List are garbage dumps. Modern landfills are not environmentally benign, but technical improvements made in landfill design and operation, combined with today's stricter standards, have lessened their negative environmental impacts and helped to renew public confidence in this waste disposal option.

THE BASICS OF LANDFILLING

Before a landfill is constructed an environmentally appropriate and politically acceptable site must be selected. Federal regulations help to guide communities in this process by creating specific require-

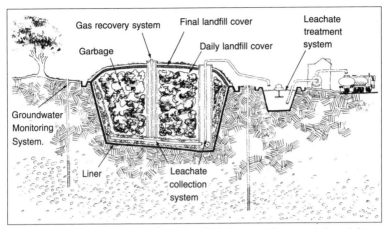

Figure 15. Landfill with Pollution Control Equipment. *Source:* Adapted from National Solid Waste Management Association.

ments to improve the siting, design, and operation of new landfills. Modern landfills are safer, but siting them is an increasingly difficult challenge. The technical constraints and regulatory standards are big hurdles in themselves—and the political issues involved are particularly volatile.

RCRA's Subtitle D restricts the siting of landfills in floodplains, wetlands, earthquake prone areas, near airports, or where the ground cannot support the weight of a landfill. State and local governments can influence the selection of a site by requiring building permits, regulating the landfill size, and enforcing local zoning ordinances. Landfill design criteria also are outlined in the new regulations. The primary focus is the liner that separates garbage from groundwater.

Modern landfills begin as a hole dug in the ground and lined with a layer of compacted soil, with a synthetic liner, or both, in which garbage is dumped and at the end of each day covered with a layer of soil, foam, or plastic. By the time a landfill reaches capacity it is a hill. Some communities have built parks, airports, and even ski slopes on top of their closed landfills. New landfills are designed to reduce the potential risks of groundwater contamination and methane emissions. To ensure better pollution control, landfill regulations require that all new and expanded landfills be constructed with a liner, a leachate collection system, a landfill gas control system, and a groundwater monitoring system. Once a landfill is full, the owner/operator must implement the landfill's closure plan previously submitted to the state. Subtitle D requires that closure plans include activities to properly close the landfill and monitor and care for it for at least 30 years after closure. All closure plans must include capping the landfill with soil.

HOW GARBAGE DECOMPOSES

Much has been written in the popular press about how garbage does not decompose in a landfill. But it does, eventually, and this is how it happens.

Aerobic Stage: Garbage put in a landfill begins to decompose with the help of oxygen. Aerobic bacteria produce water, carbon dioxide, nitrates, partially degraded organic material, and heat (often 122–158°F). This stage lasts about 2 weeks, until the oxygen is depleted.

Acid Anaerobic Stage: After the oxygen is gone, garbage continues to decompose. Anaerobic bacteria produce carbon dioxide and partially degraded organic material, particularly organic acids. Heat production is reduced. This stage lasts 1–2 years.

Methanogenic Anaerobic Stage: Methane gas is formed as a product of anaerobic decomposition. Methane and carbon dioxide are the dominant chemicals produced. This stage can last several years or decades depending upon landfill conditions (temperature, soil permeability, and water levels). During this stage methane gas can be recovered.

ENVIRONMENTAL AND HEALTH CONCERNS AND TECHNICAL SOLUTIONS

Environmental concerns surrounding landfills persist, despite substantial improvements made in landfill design and technology. Modern landfills are designed to entomb garbage and keep pollutants away from the environment, thus prolonging the decomposition process. In fact, landfills are more efficient at preserving garbage than ever before. Studies reveal that after 10–20 years in a landfill, garbage, even organic waste such as paper and food, often retains its original weight, volume, and form.[10] One 1952 newspaper found in a landfill after 35 years was readable from cover to cover.

The slow rate of decomposition does not necessarily mean that environmental risks are abated; releases of potentially toxic substances from landfills still can enter water sources and the atmosphere. Groundwater and surface water can be contaminated by leachate (a liquid generated from rain, snow, and decomposing trash that passes through a landfill, picking up contaminants along the way). Without proper pollution control equipment, potentially toxic gases and methane can be released into the air.

Landfills that meet federal standards can lessen these environmental risks. Subtitle D of RCRA requires landfill owners/operators to monitor water sources and to maintain a leachate collection system and a final cover (a 4-foot cap of soil or a combination of soil and synthetic material) for a minimum of 30 years after the landfill closes. This 30-year maintenance requirement is critical, since even the most well designed landfills will eventually fail due to the natural deterioration of liners, leachate collection systems, and final cover materials.

Air Emissions

The primary gases emitted from landfills are carbon dioxide and methane. Methane, representing 50–60 percent of landfill gas emis-

1970

▼

The U.S. Environmental Protection Agency is created.

The first Earth Day is celebrated on April 22.

The Resource Recovery Act amends the Solid Waste Disposal Act and requires the federal government to issue waste disposal guidance.

sions, is an odorless, explosive gas that is produced as organic matter decomposes under anaerobic (airless) conditions. It poses a health risk because of its explosive nature when confined in pipes, drains, and buildings. If released directly into the atmosphere, methane, a greenhouse gas, contributes to global warming and air pollution. Landfill gas also contains small quantities of other volatile organic compounds (VOCs), such as benzene and vinyl chloride. VOCs can be toxic and carcinogenic and have been tied to low-level ozone formation (smog).

Air emissions can be controlled by gas collection systems installed at new or closed landfills. Collected gas can be flared, used as fuel, or converted to electricity. Some states require gas collection. Methane monitoring is required under Subtitle D. Any excessive buildup of methane must be remediated, often by collecting and flaring the gas. Because landfill gas contains VOCs, EPA plans to regulate it under the Clean Air Act. Under the proposal, landfill operators would be required to use at least a collection and flaring system that would destroy about 98 percent of the VOCs and methane. It should be noted that burning methane generates carbon dioxide, a much less potent greenhouse gas.

Leachate

Leachate can contain a broad range of chemicals, including metals such as lead, cadmium, and mercury, and organic chemicals such as benzene. The amount of leachate generated by a landfill is influenced by precipitation levels, site topography, facility design and operation, and the type of final vegetative cover.

To lower environmental and health risks posed by leachate, landfills that receive waste after October 9, 1993 must install a groundwater monitoring system and set aside money to pay for groundwater cleanup. This should improve the operation of many landfills; of the estimated 5,800 operating landfills fewer than 1,500 had groundwater monitoring systems and only about 900 had liners.

A groundwater monitoring system consists of a series of wells located near the landfill. By sampling the groundwater around the landfill, the presence and the migration of contamination can be detected so that pollution of groundwater sources can be prevented. Managing leachate to protect the groundwater is one

Figure 16. Landfill-maintenance machines called "landfill crawlers" are shown here spreading and compacting waste at a municipal solid waste landfill. (Courtesy of the National Solid Waste Management Association.)

1971

▼

Oregon passes the nation's first
bottle bill.

of the most important issues communities must address in the design, operation, and long-term care of landfills.

Liners can be made of compacted soils or synthetic materials with low permeability or both. Recompacted clay is the preferred soil liner. Other soils that can act as a liner include other types of clay (e.g., weathered clay) and silt. Synthetic liners such as high density polyethylene (HDPE) or polyvinyl chloride (PVC) plastics also can be used. A liner that combines an engineered soil layer with a synthetic liner is called a composite liner. Under the new federal landfill regulations most new and expanded landfills will be required to use this type of liner—at least 2 feet of compacted clay and a 60-mil-thick HDPE liner (the same material used for plastic milk cartons, but as thick as about 60 garbage bags) or a 30-mil-thick liner made from other types of approved synthetic materials. The advantage of a composite liner is that each component has different resistance properties.

A leachate collection system usually consists of a system of perforated pipes built above the liner and sloped to drain to a central pump. Once leachate is collected, the most common method of disposal is to discharge the leachate to a municipal sewage treatment facility. This process must be closely monitored to avoid overloading the sewage facility and thereby threatening nearby waterways with waste overflow.

A second option is on-site treatment of the leachate. Due to the high chemical concentrations in leachate compared to municipal and some industrial wastewater, these facilities must be carefully designed and monitored. Popular on-site methods include the use of a pond to treat the leachate biologically or an on-site wastewater treatment plant.

Another treatment method is redistributing the leachate over the top of the landfill. More research on this method is needed. Leachate recirculation proponents argue that the process shortens the decomposition time of garbage, thus opening reusable landfill space, and that the wetter conditions create an optimal environment for methane production, collection, and use. Opponents claim that because recirculation fosters waste decomposition, the leachate has a higher concentration of toxics. Critics also fear that the increase in the amount of leachate generated will increase the chance of leaks through holes in the liner or the leachate collection system. They also contend that the reactions of chemical compounds created during the process have not been studied thoroughly.

ECONOMIC CONSIDERATIONS

The main objective of RCRA's landfill provision is the "protection of human health and the environment." To achieve this goal, a landfill management plan must include steps to properly monitor the landfill while it is operating and to close and maintain the facility for at least 30 years after it is closed. In addition, landfill owners must show that they have the means to finance closure and post-closure care of the landfill as well as the cleanup of any leaks.

Landfills, like WTE plants, have a built-in source of revenue. They produce methane, which is an energy source similar to natural gas, but with about one-half the heating value. Some larger landfills find it is economically feasible to collect and process methane gas. In 1992, over 100 facilities recovered methane.

A high quality collection system working under optimal conditions can recover up to 80 percent of the methane produced by a landfill. Recovered methane can either be used as a low-grade fuel or upgraded to "pipeline-quality" methane. If a market does not exist, the gas can be used at the landfill in boilers, furnaces, or electric generators. As with other natural gases, its economic value is dependent upon current market prices, proximity to markets, and the availability of federal tax credits. The revenue earned through the sale of the gas, however, by no means offsets landfill costs.

LANDFILL MINER FORTY-NINER

In order to increase energy production, generate revenue, and extend the life of a landfill, a handful of communities are turning to landfill mining. Landfill mining or landfill reclamation is the practice of digging up old refuse and dirt and reusing the excavated material. Plastics, metals, and glass can be recycled. Soil-like material can be reused as topsoil, compost, or daily cover on the landfill. Combustible landfill material can be burned in a waste-to-energy facility. Unearthing these materials creates new landfill space.

During the boom construction year of 1990, the Pitkin Resource Recovery Facility, located in Pitkin County, Colorado, began to screen incoming construction debris for material that could be resold. The sale of the material was profitable enough that during the subsequent slump in the construction market, the landfill operator turned to landfill mining of buried construction debris to make up for the shortfall in incoming construction debris.

Much of Pitkin County's excavated material is soil, fill material from construction sites, and wood waste that is sold as topsoil, composting material, or as road-basin aggregate. Although the costs can be high in terms of labor and specially designed equipment, the returns from the sale of the construction debris are profitable. In 1992, the facility's modest landfill mining operation yielded approximately $10,000–20,000. To keep pace with the demand for reclaimed materials, the Pitkin landfill operator plans to work on methods of speeding up the rate of decomposition for more efficient mining.

Motivated by raising revenue from the sale of additional energy at the county's waste-to-energy facility, Lancaster County, Pennsylvania, also turned to landfill min-

continued on next page

ing. The waste-to-energy facility was designed with extra capacity to allow for the growth of the county. Instead of keeping the extra capacity in reserve, the county decided to excavate waste from the county landfill and burn the mined garbage at the facility. In 1992, the resulting electricity was sold for a net weekly revenue of $7,530.

As a technique, landfill mining is in its early stages. What may seem to be a profitable solution may give rise to unforeseen problems. Factors such as the possibility of unearthing toxic materials, the creation of extra ash through the burning of excavated waste, odor and air emissions at the excavation site, and handling of waste should be taken into consideration as this option is explored.

POLICY RESPONSES TO LANDFILLS

Under RCRA Subtitle D, EPA has set the minimum standards for landfills. The 6 major provisions of the rule cover location restrictions, operating criteria, design criteria, groundwater monitoring and corrective action, closure and post-closure care, and financial assurance. By October 1993, all publicly and privately owned landfills receiving waste must comply with Subtitle D standards, excluding the financial assurance and groundwater monitoring requirements which must be met by 1994.

States are required to adopt and implement permit programs that ensure compliance with the new standards. To encourage adoption of the rules into a state's existing permit program, EPA structured the rules to allow "approved" states certain exemptions in the standards to meet state design or performance standards. If a state program is not approved by EPA then landfill owners and operators must adhere to the less flexible requirements of the federal standards. Specific criteria for approval of state landfill permit programs are outlined in EPA's State and Tribal Implementation Rule (STIR).

To implement the new federal standards, most states will have to revise their permitting regulations, although Pennsylvania and New York, for example, already have stricter regulations than required under RCRA. In states requiring major revisions in their regulations, the changes will inevitably affect future costs for building, operating, and maintaining landfills. As with incineration, state officials and landfill owners and operators must keep one step ahead of regulators' definition of "state-of-the-art" and anticipate future requirements.

HOW SAFE IS STATE-OF-THE-ART?

New landfills are often referred to as "state-of-the-art" facilities because of their engineered design and use of new and improved technology. Although a specific definition does not exist, characteristics commonly associated with a state-of-the-art landfill are a liner system, a leachate collection system, an engineered final cap, landfill gas control and recovery, gas and groundwater monitoring systems, and post-closure care and maintenance. Subtitle D rules include the use of the best technology available and long-term care.

The term "state-of-the-art" evokes unequivocal faith in current science and technology. However, even the best liners will eventually degrade, tear, or crack. Landfills are designed to accept waste for 10–40 years, and the new EPA regulations require owners to maintain and monitor their landfills for at least 30 years after closure. The questions remain. What happens after 30 years? Who will need protection then? Who will pay for cleaning up ground and surface water sources when leaks occur?

State-of-the-art landfills may not be a cure, but they are a tremendous improvement over their predecessors.

SIX

SPECIAL WASTES

Certain types of municipal solid waste are particularly challenging to waste management facilities, either because they are more toxic than most materials or because their size, shape, or other properties keep them from being handled with the rest of the trash. These "special wastes," as they are known, include used motor oil, tires, household appliances such as refrigerators, washers, and dryers (known as "white goods"), and household hazardous wastes such as oven cleaner, paint, solvents, and pesticides. No single "best" waste management option exists for dealing with any of these categories of wastes, and states and municipalities continue to seek new ways to reuse, recycle, or dispose of these materials safely.

USED MOTOR OIL

It does not take much oil to contaminate a lot of water. One gallon—a single oil change—can pollute a million gallons of water. Contaminants often found in used oil add to its toxicity. Approximately 600 million gallons of used motor oil are generated each year in the United States. Most businesses that generate used oil, such as service stations, repair shops,

1972

▼

Stockholm Conference on the Human Environment is held, the first international environmental conference.

and corporate and military motor pools, collect it for reprocessing, but this is not true of many of the 48 million Americans who change their own oil.

About 90 percent of the used oil generated annually by "do-it-yourself" oil changers is improperly managed. That amounts to about 200 million gallons (equivalent to 18 Exxon Valdez spills) dumped on the ground, poured down sewers, or thrown in the trash. Oil dumped on the ground or landfilled contaminates soil and may leach through the soil to contaminate groundwater supplies. Oil poured down sewers can pollute surface waters directly or disrupt sewage treatment plants. The U.S. Coast Guard estimates that sewage treatment plants discharge twice as much oil into coastal waters as do tanker accidents. The remaining 10 percent of do-it-yourself oil is collected for reprocessing.

The greatest challenges to halting the improper disposal of motor oil by do-it-yourselfers are communicating the hazards of such disposal and making available convenient, environmentally sound collection points. Statewide, regional, and community programs across the country continue to develop strategies to bolster used oil collection and recycling. Many provide public education and post information on used oil collections at stores that sell oil and encourage those retailers to collect used oil from their customers. Some communities have added oil to curbside trash or recyclables collections programs, and all household hazardous waste collection programs include used oil. Many programs encourage do-it-yourselfers to buy re-refined oil where it is available.

Alabama's 15-year-old "Project ROSE" (Recycling Oil Saves Energy . . . and the Environment), for example, helps establish used oil recycling programs, provides information on access to collection sites and curbside programs, and maintains a toll-free information number. In 1992, Project ROSE expected to recycle 8 million of Alabama's 17 million gallons of do-it-yourself used oil.

Major oil companies are getting into the act, as well. Mobil, Texaco, Exxon, and Ashland Oil are among those allowing do-it-yourself oil changers to dispose of their used oil without charge at participating service stations.

Terminology: Is burning recycling?

To the dismay of many recycling advocates, what is commonly referred to as oil "recycling" usually means burning it as fuel. More than 90 percent of all collected motor oil is burned as fuel in industrial, school, and residential boilers and space heaters. Before it can be burned, used oil usually must be processed to meet state or federal standards limiting the permissible level of lead and other contaminants. Processing the oil to meet existing standards may involve removal of certain contaminants or, more commonly, blending the used oil with virgin oil. In 1991, burning of used oil for fuel emitted nearly 600,000 pounds of lead, more than any other industrial process.[11]

Critics say that "recycling" more accurately refers to "re-refining" and they contend that the "recycling" programs initiated by some oil companies are actually *collection* programs for oil that is destined to be burned. The companies, they say, are doing nothing to strengthen re-refining technologies or infrastructure, which would be much closer to real recycling.

Less than 10 percent of the collected oil is re-refined, a more complex and expensive process that yields a correspondingly more valuable product. Re-refining a gallon of used oil can reclaim the same amount of pure lubricating oil that would take 42 gallons of crude oil to produce. Although a type of re-refining has been practiced for decades, methods used in the past produced large quantities of toxic by-products to produce oil that had limited uses. Today's technology is substantially different. Current technology is more environmentally sound and can produce re-refined motor oil that meets the most rigorous standards and specifications demanded of virgin oil. Today's re-refining industry using the new technology is young and small, with no more than one-half dozen companies operating in the United States in 1992.

Classification: Is used oil a hazardous waste?

One of the thorniest issues related to used oil collection is whether oil is a hazardous waste. Those on opposite sides of the question have these conflicting concerns:

- ♦ If used oil is classified as hazardous, efforts to recycle it will collapse: no one will want to risk criminal liability for handling it. Such classification may even increase the likelihood of improper disposal.
- ♦ If it is not classified as hazardous, collected used oil, which typically contains toxic constituents, will not be handled with sufficient care. Those advancing this argument point to the common practice of burning used oil, despite the likelihood of toxic emissions, as an example of inappropriate handling. They also cite the listing of various recycling facilities (including some of the outmoded used oil facilities) as priority Superfund sites as evidence that recycling processes are not automatically benign and need to be adequately regulated.

As a compromise, some have proposed a "modified listing" of used oil as a hazardous waste. Under a modified listing, used oil would not be considered "hazardous" for regulatory purposes until it leaves the collection point. This would exempt do-it-yourselfers, service stations, and municipal collection programs from hazardous waste liability, while ensuring that those who process the used oil would take adequate precautions. Proponents of this approach point to California as a model. Under California's modified listing program, which took effect in 1986, oil recycling has increased, and the state currently is a leader in used-oil recycling.

In 1992, EPA issued mandatory management standards covering used oil generators, transporters, burners, processors, and re-refiners, and determined that with these standards it is unnecessary to classify used oil as a hazardous waste. Many regard the standards as too weak to prevent pollution, however. In 1992, Congress considered, but did not pass, legislation regulating used oil. The legislation was reintroduced in 1993.

TIRES

The estimated 242 million tires discarded annually continue to test the creativity of waste management officials around the country. Almost 80 percent are landfilled, stockpiled, or illegally dumped. However, tires take up a lot of space in landfills, and air trapped in rims and the resiliency of the rubber make them buoyant, so that they work their way to the surface, breaking through the cover, and causing uneven settling within the fill.

The 2–4 billion tires that have accumulated in stockpiles or uncontrolled tire dumps can pose health and fire hazards. Rain water provides breeding grounds for potentially disease-carrying mosquitoes, and a fire in a tire pile is extremely difficult to put out. Tire material, primarily hydrocarbons, burns readily, while air in the rims provides a continuous supply of oxygen to feed the flames. Burning tire piles produce noxious air pollutants and toxic runoff—the mixture of oils produced by the burning tires and water used to combat the fire—that may pollute nearby waterways and groundwater.

In 1991, less than 7 percent of discarded tires were recycled, while 11 percent were incinerated for fuel value, and 4 percent were exported.

Recycling

"Recycling" refers to finding new uses for old whole tires or the "crumb" rubber and "reclaim" rubber made from them. Because

Figure 17. Of the 242 million tires discarded annually in the United States, many are sent to tire stockpiles and dumps. (Photo by S. Levy, Environmental Protection Agency.)

the rubber in scrap tires loses resiliency in reprocessing, it is now rarely used to manufacture new tires. Whole tires are used for erosion control on highway projects, for highway crash barriers, as artificial reefs in coastal areas, and for playground equipment. Tires may be split or punched to produce floor mats, gaskets, dock bumpers, and shoe soles.

As the term suggests, crumb rubber is produced by shredding or chopping tires up into small pieces. Crumb rubber can be incorporated into rubber products such as carpet padding and is increasingly being used in road paving materials. Two methods for incorporating rubber into asphalt pavement are under investigation in test projects around the country. Some states and the federal government have moved to permit—or require—the use of a given percentage of rubberized asphalt in routine road projects. Most test results indicate that rubberized asphalt pavements are more resilient and last longer. However, they cost more initially, which is the primary reason some road departments have been resistant to the new methods and materials. Another concern is the recyclability of the new asphalts. Typically when a road is being repaved the old asphalt is heated and mixed with new materials. Some highway departments worry that the rubberized asphalt may catch fire or produce noxious fumes when it is heated. Producers of the new materials deny this would happen, citing studies of successful recycling.

1976

The Resource Conservation and Recovery Act creates the first significant role for federal government in waste management. The law emphasizes recycling, as well as conserving energy and other resources; it also launches the nation's hazardous waste management program.

"Reclaim" rubber, a mixture of crumb rubber, oil, water, and chemicals that has been heated under pressure, can be used as a substitute for virgin rubber for some applications, such as floor mats, and as a small fraction of the rubber used in new tires.

Fuel

Used tires also are burned as fuel. With an energy value of 12,000–16,000 Btu per pound, slightly higher than coal, they are used in power plants, tire manufacturing plants, cement kilns, pulp and paper plants, and small steam generators. A few plants can handle whole tires, but most combustors require the tires to be shredded into what is called "tire derived fuel" (TDF). TDF can be further processed (at additional cost) to remove steel belts and beads if a combustion facility requires it. EPA reports that, using current technology, tire burning facilities can meet federal and state emission standards. The use of scrap tires as fuel and in asphalt paving is expected to grow in coming years.

Reused Tires and Retreads

Not counted as scrap tires are the 10 million tires reused and 33.5 million tires retreaded each year. These figures reflect a decline in reuse and retreading in recent years, due in part to the low cost of new tires. In addition, EPA attributes the decline in passenger vehicle retreads—from 23 million in 1987 to only 18.6 million in 1990—to the common misperception that retreads are unsafe. EPA estimates that twice as many tires were suitable for retreading. The number of truck tires being retreaded increased slightly between 1987 and 1990.

In 1989, EPA issued procurement guidelines encouraging the use of retread tires by government agencies. The National Tire Dealers and Retreaders Association maintains that properly retreaded tires are as safe and will last as long as new tires, and points out that warranties on quality retreads are comparable to those on new tires.

Legislation

Most states regulate some aspect of the storage, processing, hauling, and landfilling of scrap tires. Many states use taxes or surcharges on tire sales, vehicle registration fees, or landfill tipping fees to fund programs to clean up old tire dumps, research new applications for scrap tires, and administer regulations. The federal Intermodal Surface Transportation Efficiency Act (ISTEA), passed in 1991, requires states to use crumb rubber in federally funded highways beginning in 1994.

APPLIANCES

Major household appliances such as refrigerators, stoves, washers, and dryers, often called "white goods" for their traditional color, require special treatment due to their size and the toxic constituents some contain. In the past, some appliances were compacted and landfilled, while others were salvaged for scrap metal. Today, many states and municipalities are working to keep white goods out of landfills in order to save space and help meet materials recovery goals through appliance recycling. In 1992, 13 states had bans on landfill disposal of appliances.

> **1978**
>
> ▼
>
> The Supreme Court rules that garbage is protected by the Interstate Commerce Clause; therefore, New Jersey cannot ban shipments of waste from Philadelphia.

Diverting white goods from the municipal solid waste stream has become more complicated—and more urgent—as the scientific evidence against polychlorinated biphenyls (PCBs) and chlorofluorocarbons (CFCs) has mounted.

PCBs are present in the electrical capacitors of some appliances. To retrieve scrap metal for recycling, scrap dealers shred the old appliances, retrieve the metal, and dispose of the rest of the material, called "shredder fluff." In the shredding process, capacitors are crushed and PCBs leak out, contaminating the shredder fluff, which then requires special handling and disposal in expensive hazardous waste landfills or incinerators. In 1988, when EPA set limits for PCB contamination in wastes acceptable for normal disposal, some scrap dealers stopped accepting appliances with capacitors. Today, most scrap dealers require a guarantee that all capacitors have been removed, and some cities have established capacitor removal programs as a standard component of their waste management services. Some scrap dealers remove the capacitors themselves.

Chlorofluorocarbons and a related family of ozone-depleting chemicals, hydrochlorofluorocarbons (HCFCs), are used in refrigerators, freezers, air conditioners, and dehumidifiers. If released into the atmosphere, CFCs deplete the stratospheric ozone layer, which protects the earth from harmful ultraviolet radiation. CFCs and HCFCs are regulated under the Montreal Protocol, an international treaty to protect stratospheric ozone. More to the point for municipal solid waste managers, they also are regulated by federal law. The 1990 amendments of the federal Clean Air Act require the removal of CFCs and HCFCs from appliances prior to disposal, and prohibit venting the chemicals. The amendments went into effect in July 1992. Waste haulers must now be able to remove these chemicals or be able to transport old appliances safely to a place that can. Pick-up and handling procedures may have to be altered to prevent any damage to the appliance that might allow the chemicals to escape. Responsibility for ensuring that the chemicals are properly managed may fall to different players in the waste disposal chain, from hauler to scrap dealer or landfill operator, depending on the local situation.

By 1992, some enterprising companies had developed the technology to safely extract CFCs and HCFCs from old appliances, and other companies are expected to follow. One company in particular has received attention for its pioneering work in appliance recycling. Appliance Recycling Centers of America (ARCA), based in Minneapolis, started out refurbishing old refrigerators in 1976 but today functions as an environmental services company. The company has 6 plants, including 3 that opened in 1992.

Each ARCA plant was built in cooperation with a utility company seeking to establish an appliance take-back or buy-back program. Often, after buying a new refrigerator or freezer, people will keep the old, energy-sapping appliance for occasional use (and for lack of ready alternatives). These inefficient second appliances place a drain on the utility's energy resources. By buying the old appliance and taking it out of service, the utility can actually save money on energy production costs. The appliances are picked up and delivered to ARCA, which receives a fee for each one.

ARCA extracts CFCs, HCFCs, mercury switches, and capacitors containing PCBs from the old appliances. About 10 percent of the appliances are refurbished and the CFCs and HCFCs are recycled. The unsalvageable appliances are shredded for scrap metal, and the leftover residue is safe for disposal in an ordinary landfill.

HOUSEHOLD HAZARDOUS WASTES

Household hazardous wastes (HHW) constitute a small but potent segment of the municipal solid waste stream. In general, the term includes household products with flammable, corrosive, reactive, or toxic constituents, once the products have outlived their usefulness and enter the waste stream.

Such items as household batteries and car batteries, drain cleaners, oven cleaners, pesticides, paint, solvents, aerosol cans, motor oil, and antifreeze can contaminate air, soil, or groundwater if landfilled, or produce hazardous emissions or toxic ash if incinerated. Sanitation workers risk injury from exposure to toxic substances, or even explosions, if certain HHW buried in normal household trash are run through the garbage truck grinders and compactors or other equipment used in solid waste collection and processing. As a result, cities and states have started programs to keep these wastes out of the household trash. HHW programs generally include education, source reduction, and collections components.

The first step is getting the word out—about what household hazardous wastes are, why they should be kept out of the household trash, and where people can take them instead. Programs often encourage people to buy only what they need for a particular job or will use in a short period of time (an important exception to the "buy in bulk" strategy for source reduction) and to seek less toxic alternatives. HHW collection programs often are advertised in newspapers, on local radio and television stations, or through flyers posted at libraries and stores. (Two excellent videos, *Cleaning Up Toxics at Home* and *Cleaning Up Toxics in Business,* are available from the League of Women Voters of California; see the Resources section.)

Most household hazardous wastes collections are single-day events that allow residents to bring their hazardous wastes to a temporary drop-off site. The number of HHW collection days taking place around the country grew from 2 in 1980 to 802 in

1991. The number of permanent HHW collection facilities has grown, as well, from 1 in 1980 to 96 in 1991.

Unfortunately, HHW programs typically have very low participation rates and are very expensive to run. The chief expense is the cost of disposal at hazardous waste landfills or hazardous waste incinerators, although other waste management options such as recycling and reuse, appropriate for much of the collected materials, have costs as well. The contractor operating the program must have trained personnel, adequate equipment, and insurance, and these costs will, of course, be reflected in the contractor's fee. Since household hazardous wastes make up only about 1 percent of the municipal solid waste stream, some question whether the benefits justify the costs.

GETTING A CHARGE OUT OF GARBAGE

What began as a pilot project for button cell batteries has grown into a full-scale battery collection program in Hennepin County, Minnesota. The program collects an assortment of button cell batteries, the tiny round batteries used in watches, hearing aids, cameras, and calculators, and cylindrical AAA, AA, C, and D batteries used in flashlights, toys, radios, etc. There are approximately 550 drop-off collection centers for button cell batteries and 160–170 centers for mixed cylindrical batteries located throughout the county in retail malls, libraries, senior citizen centers, city halls, and jewelry and camera stores. The county also operates a household hazardous waste collection center that accepts household batteries.

After consolidating the batteries from the collection centers, they are sorted by chemical type and sent to one of three destinations. The mercuric-oxide, nickel-cadmium, lead-acid, and silver-oxide batteries are recycled at a metals reclamation center that recovers usable metals. Bat-

continued on next page

teries containing lithium are sent to a hazardous waste
incinerator, because no recycling options are available.
Alkaline and carbon-zinc batteries are sent to a hazardous
waste landfill. The mercury content in these batteries is
too low to profitably recover and the cost of purifying the
other materials is currently higher than the cost of virgin
materials.

Hennepin County is pleased with the program's collection rates of 70 tons in 1991 and 110 tons in 1992, and
with its success in minimizing heavy metals burned in the
county's waste-to-energy facility.

Those who have established programs say yes, while ac-
knowledging the difficulty of quantifying the benefits. They argue
that the cost of cleaning up contamination is far greater than
preventing it in the first place. Also, program managers have found
that residents who do participate bring in substantial quantities—
the average is 100 pounds of HHW per participant household.
Moreover, HHW programs help make people more knowledgeable
about potential exposure to chemicals at home. Finally, regardless
of perceived benefits and costs to the community, some type of an
HHW program may be necessary if state regulations or the local
nonhazardous landfill or incinerator operator prohibit disposal of
certain hazardous materials.

Another potential incentive to establish HHW programs is
proposed legislation exempting communities from certain Superfund liability. Under current law, an industry charged with the cost
of cleaning up a declared Superfund site may sue other parties that
used the site to help with cleanup costs. This includes municipalities and others that had used the facility for municipal solid
waste disposal. Legislation proposed in 1993 would protect parties
using a facility for municipal garbage disposal from such industry
lawsuits. It would also limit a municipality's liability for cleanup
costs in settlements with the federal government, if the city met

1979
▼

EPA issues landfill criteria that
prohibit open dumping.

certain provisions. One of the central provisions is that the city has an established household hazardous waste collection program within a given period after the bill's passage.

To achieve the goal of significantly diminishing the chances of contamination from improper HHW disposal, participation rates in HHW programs must be increased. So far, however, most programs have been unable to bring down costs per participant so that greater participation means higher costs for handling and waste management. Before program managers can set goals and push to increase participation, they must be sure they can afford the increased costs that are likely to result. Most HHW programs are funded through dedicated sources—taxes or surcharges imposed specifically to finance the program. Frequently these include surcharges on tipping fees or municipal solid waste collection fees and surcharges on wastewater treatment taxes or fees.

To increase effectiveness, communities have expanded their public education programs to reach a broader spectrum of the population and enlist greater support from public agencies. To make dropping off HHW more convenient, some have been able to increase the frequency of their one-day events and some have established permanent facilities. Some communities collect specific items such as paint and household batteries at the curb. No community has separate curbside pickup for the full range of household hazardous wastes, however, because of health and safety concerns about certain types of waste being left unattended at the curb.

One objection to HHW programs has been that the amount of hazardous wastes generated by households is insignificant compared to that generated by the "conditionally exempt small quantity generator" (CESQG). These are businesses exempted from federal hazardous waste mandates under the Resource Conservation and Recovery Act (RCRA) because they do not produce a large quantity of waste—less than 100 kilograms per month. Some observers have wondered how worthwhile it is to divert relatively insignificant amounts of hazardous wastes from households, when, under federal law, the hazardous wastes from small quantity generators legally can be disposed at ordinary, nonhazardous landfills and incinerators. (A few states and municipalities regulate CESQG-generated hazardous wastes; most do not.) EPA is encouraging the collection of these wastes at HHW programs, even though RCRA does not require it.

Figure 18. (Courtesy of S. Levy, Environmental Protection Association.)

SEVEN

SITING FACILITIES

As discussed throughout this handbook, managing our municipal solid waste requires integrating source reduction, recycling, composting, and safe waste disposal methods. And to do this, the construction of new waste facilities is often required to increase a community's waste disposal capacity, handle collected recyclables, improve technology for environmental protection, or meet other goals. Whether a community decides to build a materials recovery facility, composting facility, incinerator, or landfill, choosing the site is not an easy task.

In the early 1970s, between 300–400 municipal landfills were built a year in the United States; in the 1980s, the numbers dropped to 50–200. The number of new waste-to-energy (WTE) plants sited also dropped in the 1980s to an average of 12 each year. In 1992, only 2 WTE plants were built. Public opinion polls show that most people oppose the siting of new waste facilities in their community. Therefore, it is no surprise that siting a facility can take 5–8 years.

The main reasons for public opposition are that people living near a proposed site often perceive the siting process as unfair, and they lack confidence that the proposed facility will adequately protect human health and safety and the environ-

1984

▼

During the Olympic Games in Los Angeles, athletes, trainers, coaches, and spectators produce 6.5 million pounds of trash in 22 days, more than 6 lb. per person per day (vs. the national average of 3.6 lbs. per day).

ment. To describe this growing opposition, an array of acronyms has emerged, ranging from the popular term NIMBY (Not In My Backyard), LULU (Locally Unwanted Land Use), and NIMTOF (Not In My Term Of Office) to BANANA (Build Absolutely Nothing Anywhere Near Anyone). Although these acronyms symbolize the attitude of opposition groups, it is too easy to fault public opposition for the failure in siting waste facilities. Local resistance is only one of many obstacles to overcome.

One way to try to address public concerns about the potential health and environmental risks associated with a specific facility is through the use of risk assessments. These technical studies attempt to quantify the severity and likelihood of harm to the local population's health and the environment. Typically, risk assessments are used to evaluate the risks within a single management method (e.g., mass burn waste-to-energy plant). By assessing a particular method, the pollutants of greatest concern can be pinpointed so the proper steps can be taken to control them. Using risk assessments to compare different management methods is more difficult because of the differences in the number and type of pollutants emitted, potential pathways of exposure, potential effects, and facility designs. However, technical studies often do not provide definitive answers, so although risk assessments can be used to address local concerns, they do not always eliminate public opposition.

Siting a landfill, incinerator, or materials recovery facility requires a well planned, comprehensive process that addresses public concerns and perceived risks, economic and environmental impacts, and political and social issues. The Environmental Protection Agency and numerous groups have examined siting issues and developed siting guidelines (see the *Resources* section). Three useful resources are the *Keystone Siting Process Handbook* and *Keystone Siting Process Training Manual* written by participants in workshops conducted by the Keystone Center, and the "Facility Siting Credo" developed during a national workshop sponsored by the Massachusetts Institute of Technology and the University of Pennsylvania's Wharton School of Business. The guidelines that follow are based on these resources. They are designed to help citizens work with local government to select a waste facility that meets the needs of the community and an appropriate site. The step-by-step process also can be used to evaluate and improve a siting process that is already under way. In addition, observers can use the guidelines to determine if the siting process is well structured and addresses important issues.

GUIDELINES FOR FACILITY SITING

Step One: Demonstrate the need for the facility.

A waste facility must address a real need in the community. If people agree that a facility is needed, a foundation to work from can be established.

- Determine the current and future waste capacities of the community's existing landfill, incinerator, and recycling, composting, and source reduction programs.
- Analyze the municipal waste stream—what is in it, how much is there, who generates the waste?
- What are the current and projected disposal costs?
- What would disposal costs be if the proposed facility was not built? For example, if a landfill is to close in 2 years, is it possible to ship the garbage elsewhere, and how much would it cost?
- What risks do current waste facilities pose? How will these risks change if a new waste facility is built?

◆ Is there anything that can be done to divert materials from existing facilities? For example, can yard wastes be composted or certain construction materials be reused instead of being sent to a landfill?

Step Two: Build a public participation process.

Participation means being informed *and* involved. Early, substantive, and continual public participation creates a working dialogue between the facility developer and the community. Establishing an effective public participation process costs time and money, but not involving the public is significantly more costly in the long run. Dialogue is cheaper and more productive than litigation.

◆ Identify the "public" (e.g., neighborhood coalitions, residents near the proposed site, environmental and public interest groups, industry, educators, businesses, taxpayers). Of course the "public" is a diverse and dynamic group. What draws them together is proximity to or use of the proposed facility, economic impact, social and environmental issues, and legal mandates.

◆ Provide information on the proposed project.

◆ Promote the exchange of information between the public and other parties in the siting process.

◆ Provide an active role for groups (e.g., low-income and minority) affected by the decision but often left out of the decision-making process. This might require that the facility developer subsidize some groups' costs for research, travel, etc.

◆ Give all the affected groups the opportunity to be at the bargaining table. Work to address public concerns.

Step Three: Build trust in the siting process.

Decision makers can no longer use the traditional decide-announce-defend strategy for siting facilities, keeping the decision-making responsibility in the hands of only a few individuals. Today, the siting process must:

- ◆ Include all the stakeholders, including the public.
- ◆ Be fair and impartial.
- ◆ Build trust in the site selection process, in the technology, management, and operation of the facility, and in mitigation strategies.
- ◆ Provide an honest assessment of the positive and negative impacts of the facility.

Steps One and Two are essential to developing public trust.

1986

Rhode Island enacts the nation's first statewide mandatory recycling law. Citizens and businesses must separate recyclables from their trash.

Fresh Kills on Staten Island, New York, becomes the largest landfill in the world.

The Blue Box recycling container first appeared on the curbside of Ontario, Canada, under the Ontario Multi-Material Recycling, Inc. program.

Step Four: Evaluate environmental and health concerns.

- ◆ Perform an open assessment of the environmental and health risks of the proposed facility. This will help to build trust among the public and possible opposition groups as well as help to select an environmentally appropriate site.
- ◆ Know the regulations that apply to the facility and the permitted discharge/emissions allowances.
- ◆ Determine the probability of accidental releases of pollution from the facility into the environment as well as routine emissions.
- ◆ Determine appropriate pollution control equipment and management practices.
- ◆ Consider the proximity of the proposed facility to people, groundwater and surface water, and fragile ecological systems. For example, if a facility is sited near a school, study

the potential health risks to children and whether preventive measures should be taken or another site selected.

Step Five: Evaluate the economic impacts.

Both the community and the facility developer assume certain expectations will be met. The community expects that the economic and environmental costs to the community will be minimal. The developer and facility operator expect to earn a profit on their investment. If there is little chance of fulfilling the expectations of either party then the project will fail.

- ◆ Make it clear who owns and who operates the facility. Determine in advance who is responsible for what.
- ◆ Calculate the costs to build and operate the facility.
- ◆ Estimate the costs to the community (e.g., changes in property values and possible impacts on local industry).
- ◆ Determine if the community wishes to be compensated for hosting the facility. If so, how?
- ◆ Determine if the developer and operator have adequate resources to operate long term and provide financial assurance for the life of the facility and post-closure costs.

Step Six: Mitigate the impacts of the facility.

To reduce the risk of economic or environmental costs to a community, mitigation measures can be negotiated between the facility developer and the community.

- ◆ Require technology to avoid specific environmental or health problems (e.g., incinerator scrubber system or landfill groundwater monitoring system).
- ◆ Consider the replacing of environmentally sensitive areas as a mitigation measure, such as the construction of an artificial wetland to replace the destruction of the original wetland. This example illustrates how difficult it is to

guarantee that agreed upon impacts will be avoided or reduced—the wetland will be replaced, but there is no assurance that the artificial wetland can replicate all the complex ecological functions of its natural counterpart. This raises the question, if mitigation measures fail will the community be compensated?

♦ If impacts cannot be mitigated, then consider compensation. Many communities are having to make these difficult decisions as facility developers offer an array of economic benefits—host fees (payments to the community from the developer), agreements to employ local residents, free trash disposal, and so on. Compensation is controversial because it does not lessen the potential damages, it only provides something in exchange.

♦ Consider the issue of financial compensation to offset potential damages. Types of compensation include guaranteeing the property value of houses near the proposed site, providing funds for services that are associated with the facility, such as additional fire-fighting equipment or road improvement in the area, or paying for community facilities, such as a community center or addition to the local school.

♦ Consider the trading of land as a form of compensation. For example, a community might allow a developer to build on a section of parkland if the developer agrees to give the community a parcel of undeveloped land with ecological value (e.g., land for a wildlife sanctuary).

♦ Address community concern and perceived risks of traffic, noise, odors, etc.

Step Seven: Evaluate the social impacts.

Site selection based solely on scientific and technical assessments does not adequately address many social concerns. Local government also should:

♦ Address equity in the selection of a site. A neighborhood that is already host to an undesired facility such as a prison, power plant, or waste water treatment plant may feel that it has its share of "Locally Unwanted Land Uses" (LULUs) and will not want to accept another one. Often this is difficult to avoid because communities tend to site industrial-type and other noxious facilities together, and zoning ordinances may severely limit the options for siting such facilities.

♦ Address "environmental equity," the equitable distribution of environmental risks among racial and socioeconomic groups. Studies have documented the inordinately high number of undesirable facilities sited in minority and low-income communities.[12] Explanations for such siting patterns range from the facility's proximity to other industries, to the cost of land, to political clout, to racial discrimination.

♦ Discuss the facility's impact on the community image, alternative and future land uses of the proposed site, and possible budget trade-offs among other community services, such as the transfer of funds from the fire department's budget to fund a waste facility.

♦ Evaluate other factors of public concern.

Step Eight: Acknowledge the political nature of the process.

Siting a waste facility is as much a political as it is a technical issue. The nature of the siting process alone lends itself to a politicized situation. The parties involved—which could include elected officials, a solid waste commission comprised of elected officials, the public, and possibly a private developer—all have a vested interest in a particular outcome and will inevitably use their political clout to obtain their goal. Siting is ultimately a decision of "who gets what."

To avoid an "us versus them" scenario, those managing the siting process should:

- Allow equal access for the public (supporters and opponents) to information and provide it with a substantive role in the decision-making process.

- Use a neutral party to facilitate meetings between citizen groups and the local government or private waste company.

- Use technical experts or scientists to set environmental or health criteria. The developer and the public should agree upon the criteria to be used.

- Hold workshops to educate the public on legal and technical issues before public meetings. Provide technical support for citizens.

- In addition to public meetings, encourage interaction between the developer and the public during the siting, design, and permitting process, as well as the eventual operation of the facility.

- Be willing to negotiate to create a win-win situation.

In addition to participating in the siting process, citizens sometimes can be involved in the permitting and review process. The National Environmental Policy Act (NEPA) requires federal agencies to prepare an environmental impact assessment (EIS) before starting a project that will have a "significant" effect on the environment. An EIS evaluates a project's unavoidable environmental effects and the range of alternatives including a "no action"

1987

The Islip, Long Island, garbage barge is rejected by six states and three countries, drawing public attention to the landfill capacity shortage in the Northeast. The garbage is finally incinerated in Brooklyn and the ash brought to a landfill near Islip.

Commission on Racial Justice, United Church of Christ publishes *"Toxic Waste and Race in the United States,"* the first report to comprehensively document the presence of hazardous wastes in racial and ethnic communities.

scenario. (EPA-issued permits are exempt from NEPA requirements because the agency's permitting process is considered comparable to NEPA's). Once a draft EIS is completed it is made available to interest groups and individual citizens for comment. The agency issuing the EIS must respond to each criticism raised during the review of the draft. Critics may question the accuracy of the facts, the procedures used to forecast impacts, or the omission of specific environmental effects. A commenter also may request that another alternative be studied.

If your community's project does not fall within the criteria for a NEPA review, then check your state and local permitting laws. Many states have passed statutes modeled on NEPA and often the scope of these "little NEPAs" is broader. The criteria for conducting a state EIS vary. Some states require an EIS for any facility that will emit more than a certain amount of pollutants; other states require a state permit for every facility.

The benefits of the NEPA process are that project developers are required to carefully consider environmental issues and citizens are given the opportunity to comment on the EIS. As a result, the process can improve a project that has moved past the siting process.

Figure 19. Ways To Participate in the Siting Process

Techniques	Features	Advantages	Disadvantages
Public meetings	Less formal meetings for people to present positions, ask questions, etc.	A forum that allows the public to be heard on issues. Can be structured to permit small group interaction—anyone can speak.	Unless small-group discussion format is used, permits only limited dialogue. May get exaggerated positions or grandstanding. Requires staff time/preparation.
Public workshops	Smaller meetings designed to complete a task.	Very useful for tasks such as identifying project criteria or evaluating sites. Encourages dialogue and good for consensus-building.	Limitations on size may require several workshops in different locations. Is inappropriate for large audiences.
Advisory committees/task forces	A group of representatives from major interest groups and representatives of the affected local community that act as a policy, technical, or citizen advisory group.	Provides oversight of the project. Promotes communication between key constituencies. Anticipates public reaction to publications or decisions. Provides a forum for reaching consensus on issues.	Controversy can arise if "advisory" recommendations are not followed. Requires substantial commitment of staff time to provide support to committees.
Local review committees	Group of citizens from the community and surrounding localities appointed to develop a report outlining local concerns and how the facility applicant addresses those concerns.	Provides a means for dialogue between the applicant and the citizens. Addresses nontechnical issues. Conflicts are identified and resolved. Provides reliable information to the community.	Limited to reviewing one facility at one proposed site—it does not explore alternatives.

Figure 19. Ways To Participate in the Siting Process (*continued*)

Techniques	Features	Advantages	Disadvantages
Focus groups	Small discussion groups established to give "typical" reactions of the public. Conducted by professional facilitator. Several sessions may be conducted with different groups.	Provides in-depth reaction to publications, ideas, or decisions. Helps to predict emotional reactions.	Can get reactions, but no knowledge of how many people share those reactions. Might be perceived as an effort to manipulate the public.
Interviews	Face-to-face interviews with key officials, interest group leaders, or key individuals.	Can be used to anticipate issues or anticipate the reactions of groups to a decision. Provides a way to assess "how we are doing."	Requires extensive staff time.
Hearings	Formal meetings where people present speeches and presentations.	May be used as a "wrap-up meeting" before a final decision. Useful in preparing a formal public record for legal purposes.	Does not permit dialogue. Requires time to organize and conduct.
Hotlines	Publicized phone number to field queries or provide central source of information about the project.	Provides people with a direct contact. Provides a one-stop source of information.	Is only as effective as the person answering the hotline phone.
Public tours and walk-throughs	Organized tours of solid waste facilities or programs.	Provides hands on experience. Provides the opportunity to ask workers and owners questions directly.	Is inappropriate for large groups.

Techniques	Features	Advantages	Disadvantages
Public opinion surveys	Carefully designed questions are asked of a selected representative of the public.	Provides a quantitative estimate of general public opinion.	Provides a "snapshot" of public opinion at one point in time—opinion may change. Assumes all viewpoints count equally in decision. Costs money and must be professionally designed.
Citizen monitoring committees	A group of citizens to monitor ongoing operations of a facility.	Maintains public involvement after the facility is up and running. Provides assurances to the community that the facility is meeting local, state, and federal standards.	Difficult to sustain public involvement over time. May be hard to obtain sufficient company information. Can be perceived as confrontational or adversarial.
Plebiscite	Citywide election to decide where or whether a facility should be built.	Provides a definite, usually binding, decision on where or whether a facility should be built.	"Campaign" is expensive and time-consuming. People may be flooded with information from the side with the largest "war chest."

Source: Adapted from EPA, *Sites for our Solid Waste: A Guidebook for Effective Public Involvement.*

Figure 20. Ways To Share Information with Others

Techniques	Features	Advantages	Disadvantages
Briefings	Personal visit or phone call to key officials or group leaders to announce a decision, provide background information, or answer questions.	Provides background information. Determines reactions before an issue "goes public." Alerts key people to issues that may affect them.	Requires time.
Presentations to civic and technical groups	Deliver presentations to key community groups.	Stimulates communication with key community groups. Can provide in-depth feedback.	Some groups may be hostile.
Fact sheets	Brief summary of the key facts.	Simple way to inform the public of the issue. Useful reference material.	Requires staff time and costs money to prepare, print, and mail.
Newsletters	Brief description of what is going on in the project usually issued at key junctures to all people who have shown an interest in the project.	Provides more information than can be presented through the media to those people who are most interested. Provides information prior to public meetings or key decision points. Helps to maintain visibility during extended projects.	Requires staff time and costs money to prepare, print, and mail. Stories must be objective and credible so people will not consider the newsletters to be propaganda.
Mailing of reports	Mailing of report/study to other agencies and leaders of organizations and interest groups.	Provides full and detailed information to people who are most interested. Can increase the credibility of the study and provides an opportunity for comments.	Costs money to print and mail. Some people may not read the reports.

Techniques	Features	Advantages	Disadvantages
News releases	A short announcement/news story issued to the media to get interest in media coverage of the story.	May stimulate interest from the media. Useful for announcing meetings or major decisions.	May be ignored or not read. Cannot control how the information is used.
Press kits	A packet of information distributed to reporters.	Stimulates media interest in the project. Provides background information which reporters use for future stories.	May be ignored or not read. Cannot control how the information is used.
News conferences	Brief presentation to reporters, followed by question-and-answer period and often accompanied by handouts of the group's comments.	Stimulates media interest in the project. Direct quotes often appear in television and radio. May draw attention to an announcement or generate interest in public meetings.	Reporters will only come if the announcement/presentation is newsworthy. Cannot control how the information is presented.
Feature stories	In-depth story in newspapers or on radio and television.	Provides detailed information to stimulate interest in the project. Can stimulate interest prior to public meetings.	Newspaper will present the story as editor sees fit.

Figure 20. Ways to Share Information with Others *(continued)*

Techniques	Features	Advantages	Disadvantages
Newspaper inserts	Much like a newsletter, but distributed as an insert in a newspaper.	Reaches the entire community with important information such as project need and sites being considered. Is one of the few mechanisms through which parties can tell the story their way.	Requires staff time to prepare inserts and money to distribute. Must be prepared to newspaper's layout specifications.
Paid advertisements	Purchased advertising space in newspapers or on radio and television.	Effective for announcing meetings or key decisions. Story presented the way you want.	Advertising space can be costly. Production costs for radio and TV ads can be expensive.
Public service announcements (PSAs)	Short announcement provided free of charge by radio and television stations as part of their public service obligations.	Useful for making announcements such as for public meetings.	Many organizations compete for the same space. Story may not be aired or may be aired at hours when there are few listeners.
Information centers	An information desk located in a public building such as a public library or the town hall.	Accessible to everyone in the community. Efficient and inexpensive way to distribute information.	May reach a limited number of citizens. Requires staff time to prepare and/or collect materials and to keep the table stocked with information.

Source: Adapted from EPA, *Sites for our Solid Waste: A Guidebook for Effective Public Involvement.*

EIGHT

FINANCING SOLID WASTE FACILITIES AND PROGRAMS

The mounting cost of municipal solid waste management comes at a time when state and local governments are confronting tighter budgets and growing public demands for services. Many local governments are taking on more financial responsibility as the number of federal and state grants available for solid waste management declines. To handle this increased responsibility, local officials must be more economical and creative when funding solid waste management projects.

For many communities the operating cost of solid waste collection and disposal is the fastest growing budget item. The costs to manage our garbage can be substantial. Capital costs alone are often in the millions of dollars. Although the cost to build a waste management facility varies from region to region and facility to facility, the budget impact is significant. The average cost to build a mass burn waste-to-energy plant, for example, is about $100,000 per ton of daily capacity. Construction of a state-of-the-art landfill is about $5 per ton. (This figure accounts for only 15–25 percent of total costs because operating costs are so high, being 40–50 percent of total costs.) A materials recovery facility (MRF) with a 100 tons-per-day capacity averages $18,700 per ton.[13]

Tipping fees and processing costs also add to solid waste management costs, and they vary tremendously from region to region. The National Solid Wastes Management Association (NSWMA) reports that the average landfill tipping fee ranges from $11 per ton in the west central states to $65 per ton in the northeastern states. The average cost to process recyclables at an MRF is $50 per ton. In addition to building new waste disposal facilities, communities also must secure funds for the operation and maintenance of existing facilities and services.

FINANCE OPTIONS FOR AN MSW PROJECT

REVENUE SOURCES

Taxes:
Property
Sales
Municipal Utility
Special Tax Levies

User Fees:
Uniform Fees
Variable Can Fees
Disposal Site Fees

**Revenue From
Resource Recovery
Programs:**
Recycling
Composting
Waste-to-Energy
Landfill Methane Gas

CAPITAL SOURCES

Borrowing:
General Obligation Bonds
Municipal Revenue Bonds
Bank Loans
Leasing
Other Types of Bonds
    Short Term Tax-Exempt
    Financial Notes

**Current Revenue
Private Investment:**
Full Service Contract
Merchant Plant Contract
Turnkey Contract

Source: EPA, *Decision-Makers Guide to Solid Waste Management*

Many factors contribute to the rising cost of waste management, including wages, equipment costs, the growing volume of waste, and new federal and state regulations. To encourage public and private investment in waste management projects, many communities are creating integrated solid waste systems that provide for source reduction, recycling, composting, incineration, and landfill services.

1990

McDonald's phases out its polystyrene foam clamshell and replaces it with a paper-based wrap.

To finance waste management services, local governments can either raise the needed revenue or contract with private companies to provide waste management services. The first step is determining the full costs of the waste management plan by estimating the capital costs, the operating costs, and the projected revenue. In general, communities can generate revenue for solid waste projects through taxes, user fees, revenue from resource recovery programs, borrowing, and privatization of services.

SOURCES OF REVENUE

To operate and maintain solid waste services, a steady flow of revenue is required. Revenue is generated through three primary sources—taxes, user fees, and resource recovery programs.

Taxes

Taxes are the most common source of revenue for municipal solid waste services. Tax revenues are added to a community's general fund, and money is allocated to solid waste programs through the budget process. The disadvantage to funding waste services through local taxes is that residents do not know how much they are paying for waste services. The specific costs of garbage collection and other waste services are hidden in the total tax bill. This lack of a direct out-of-pocket expense for waste services means that residents have less incentive to practice source reduction and recycling.

Communities rely on various kinds of taxes to pay for waste management services. *Property taxes* have been frequently used, but tax reform and the impact of regional economic problems on this tax revenue source are changing the picture in many communities.

A *sales tax* levied on goods and services also can be used to support local garbage services. Communities with a high recreational or tourist trade find this tax appealing. The disadvantage is that a sales tax provides a limited and variable revenue stream compared to other types of taxes.

Another option is a *municipal utility tax* levied on privately or municipally owned utilities in a community. Some or all utilities such as telephone, electric, gas, and water companies may be required to pay the tax, which typically is passed on to the utility users. The advantages to this tax are that individual billing is not required and the tax rate can often be set by an ordinance instead of by voter referendum. However, as with the sales tax, the revenue stream of a municipal utility tax is limited and unsteady compared to other taxes.

Special tax levies are another alternative for communities, if state statute allows them. The tax rate is usually limited by statute and based on the assessed valuation of property. A citizen referendum is usually not required, but often competition for special tax revenue exists among other nonbudget services such as hospitals, parks, and civic centers.

User Fees

An increasingly popular option for financing solid waste management is through the imposition of user fees. User fees are paid by service users directly to the service provider—the municipality or private company that collects and disposes of garbage. Fees are based on the level and type of service provided.

A *uniform user fee* charges all users (residents and businesses) that receive a particular service the same fee. For example, all users receiving curbside garbage collection service are charged the same fee and all those receiving backyard garbage collection service are charged the same fee. Uniform fees are efficient and inexpensive for local government to administer if the charge is added to an existing utility or property tax bill. For example, residential water bills issued by the local government also can include garbage collection charges.

1991

Seattle, WA, starts the first known high-volume, long-distance, dedicated railhaul operation of MSW in the U.S.

A *variable user fee* or pay-per-can system charges all users (residents and businesses) according to the amount of garbage they throw out, instead of a flat fee or tax. Typically, the fee is based on garbage volume, with users putting their trash in specially marked bags or containers to assure proper billing. To establish a successful pay-per-can system, a standardized collection system is required so that garbage can be easily collected and residents and businesses charged accordingly.

Some communities combine a pay-per-can system with free pick-up of recyclables. (See Chapters 2 and 3 for further discussion.) However, recycling does carry a net cost, so communities must be prepared to increase subsidies for recycling as the initiative grows.

User fees generate revenue, raise citizen awareness of waste collection, processing, and disposal costs, and encourage recycling and other consumer behavior that supports desired public policies. The downside is that the American public is accustomed to paying only a nominal fee for garbage collection, and user fees are generally higher because they are designed to cover the full costs of the service. This initial objection to user fees sometimes can be eased through public education and participation in solid waste programs. However, communities must consider the impact of user fees on low-income residents, and deal with administrative and billing problems.

Disposal site charges or *tipping fees* are another type of user fee. These fees are usually charged to the municipality or garbage hauler when garbage is dumped at a landfill or incinerator and passed on to consumers in trash disposal bills. Tipping fees, as with other user fees, should reflect the full cost of a landfill or incinerator. The operating costs should account for daily operating costs as well as disposal costs (costs for ash disposal or landfill closure and post-closure maintenance costs), capital costs, pollution control equipment costs, and liability costs. The difficulty is that fees that reflect true costs may cause haulers to take their waste elsewhere.

Resource Recovery Programs

Revenues from recycling, composting, waste-to-energy, and methane gas recovery programs provide another source of operating revenue for solid waste management. Resource recovery projects generate revenue through the sale of recovered materials and energy, in addition to reducing the amount and cost of waste sent to landfills. Waste-to-energy (WTE) plants and landfills with methane gas recovery programs can realize savings by producing energy to use at the facility. The disadvantages of relying on revenue from recovery programs are that market prices are volatile for recyclables and the power generated at WTE plants and landfills, and buyers may be located far from the project area. Public officials and community groups can help by assisting with the development of local markets for recovery program products.

FLOW CONTROL: THE BATTLE OVER TRASH

While certain states are trying to get rid of their garbage by shipping it to other states, many are trying to solve their problems by keeping their garbage at home. In their search for ways to guarantee a steady flow of revenue to public disposal facilities, some states and municipalities are passing flow control laws that direct garbage to disposal facilities of their choice, for example, state owned landfills or incinerators. A monopoly on garbage allows a state or a municipality to eliminate competition from private and out-of-state disposal facilities, to charge higher commercial waste disposal fees, to subsidize residential disposal fees, to guarantee that the public waste facilities will receive enough garbage to operate at their full capacity, and to limit future liability (e.g., Superfund liability). This may seem like a good idea for managing municipal solid waste, but recent court decisions have cast doubt on the legality of certain uses of flow control legislation.

Flow control laws are challenged based on the commerce clause of the United States Constitution, which gives Congress the power "to regulate commerce . . . among the several states." State flow control laws are reviewed by courts based on whether they regulate state and interstate commerce equally with only incidental effects on interstate commerce, or whether they discriminate against interstate commerce either in written law or in practice.

For example, Rhode Island attempted to support its state landfill by charging full-cost fees for commercial waste and below-cost fees for residential garbage. To prevent commercial waste haulers from dumping at less expensive facilities, the state mandated that all commercial

continued on next page

waste generated in the state must go to state licensed facilities (only one such facility exists in Rhode Island and it is state owned). The U.S. District Court ruled that this law violated federal commerce laws by discriminating against interstate commerce.

Two counties in Minnesota attempted to support a compost facility by requiring that all waste generated in the counties be dumped there. The tipping fees at the compost facility in Minnesota ($70/ton) were considerably higher than at the closest private landfill in Iowa ($30/ton). Again, this legislation was ruled to discriminate against interstate commerce.

If a state law has only an incidental effect on interstate commerce, the law will violate the commerce clause only if the burden on interstate commerce is excessive in comparison to the local benefits gained. If a state law discriminates against interstate commerce, it is in violation of the commerce clause unless the state establishes that the law serves a local purpose that could not be served by a nondiscriminatory method. If a flow control law supports economic protectionism from other states, it will be ruled unconstitutional. The Rhode Island and Minnesota facilities under question did serve local purposes, but the courts ruled that they discriminated against interstate commerce and violated the commerce clause. Flow control legislation also could violate state and federal antitrust laws. Whenever a state is considering such legislation, a thorough review of the state and federal laws is required.

Despite all the legal ramifications surrounding flow control legislation, its use has only been limited, not banned, by courts. Until Congress passes federal legislation to regulate a state's ability to control the flow of garbage, states and municipalities will continue to enact flow control legislation, albeit more carefully.

SOURCES OF CAPITAL

Capital for construction of a waste facility or implementation of a solid waste program is generated through three primary sources—borrowed funds, current revenue, and private investment.

Borrowed Funds

Bonds are the most common form of capital financing. Over the past 10 years, municipalities have switched from relying on general obligation bonds to using revenue bonds and, in particular, solid waste system revenue bonds.

General obligation bonds are considered the most secure of all municipal securities because they are guaranteed by the municipality's ability to levy taxes. However, they also require the municipality to assume all project risks, while the bondholders are not held accountable for the success of the project. Many states require voter approval before general obligation bonds can be issued. General obligation bonds are practical for small and medium communities because small-scale projects can be grouped together to raise capital.

In order to share the burden of risks for major solid waste projects, municipalities also can issue *revenue bonds.* Unlike general obligation bonds, which are backed by the full faith and credit of the municipality, revenue bonds allow a municipality to limit its obligations to a specific project. A revenue bond is issued, without voter approval, to finance a single facility, and revenue from that facility is used to repay the bond. An advantage of revenue bonds is that those using the facility pay user fees for the service received and those who do not use the facility are not required to pay. A simple example is a toll road. The cost to build and maintain the road is financed by tolls paid by those driving on the road. Because revenue bonds distribute project risks among bond issuers and bondholders, they are not as flexible and have a higher interest rate than general obligation bonds.

1992
▼
The United Nations Conference on Environment and Development (UNCED), the "Earth Summit," hosted by Brazil, produces Agenda 21, the world's blueprint for sustainable development in the 21st century.

One type of revenue bond that is becoming increasingly common is a solid waste system revenue bond. A system revenue bond combines the revenue from all waste disposal services—recycling, landfill, and incineration. The advantage of this bond is that it is dynamic. Bonding authority may be given to the municipality to finance a recycling program and landfill, but if the municipality later decides to build an incinerator, financing for the incinerator may be added to the bond. A system revenue bond allows a municipality to support all aspects of a comprehensive waste disposal system without creating competition between services.

Communities also can obtain capital through a *bank loan* or *leasing agreement.* A bank loan is commonly used to purchase small-scale capital equipment such as trucks or to balance the cash flow of the solid waste system. Leasing is an alternative means for acquiring equipment and land. Both bank loans and equipment leasing agreements are short term, while land leasing agreements are typically for 5 or more years.

State and local governments also can issue *short-term tax-exempt financial notes.* These are not bonds per se, but are a type of municipal debt. The notes are issued with the expectation of receiving revenue or issuing a long-term bond in the near future. For example, a state could issue a tax anticipation note or revenue anticipation note with the expectation of receiving tax revenues or other income in the near future.

Current Revenue

A "buy as you need it" option is appropriate for purchasing capital equipment or replacing equipment. It is frequently used to purchase collection vehicles and landfill disposal systems. The advantage to this method is its simplicity since special legal and institutional arrangements are not necessary.

Private Investment

A third option for local government is to contract with a private company to raise the capital, purchase the equipment, and operate the facility. The most common approach is a *full service contract* under which the community hires a company to design, construct and operate the plant. The community specifies what type of plant should be built and the performance standards, and all other operational decisions are made by the private company. Under this approach the plant can be owned by the community, the company, or both.

A *merchant plant contract* is another financing approach. Under this contract the private company designs, builds, operates, and owns the facility. The community pays on a dollar-per-ton basis to dispose of its waste at the facility. Disposal agreements vary; often, for example, the "host" community is given a tipping fee discount.

Some facilities are run as *turnkey operations*. A plant is designed and built by a contractor to meet a community's specification and then the keys to the facility are turned over to the community or another contractor to own and operate.

Privatizing the ownership or operation of a facility can offer a community the opportunity to remove itself from the operation of a solid waste facility, saving staff time and resources. In certain instances, even if the community must pay more for the service rendered than if it provided the service directly, the trade-off in time and resources saved can be worthwhile for understaffed and overburdened government offices.

PUBLIC-PRIVATE PARTNERSHIPS HELP TO MAKE THE ENDS MEET

Constrained by limited budgets, local decision makers are looking for new and innovative ways to deliver environmental services. Public-private partnerships are one way local governments can ensure environmental protection while keeping a lid on their growing share of financial responsibility for environmental programs.

A public-private partnership for solid waste management is a contractual relationship between public and private parties. The partners agree to share the responsibility for one or more of the following activities of a facility: financing, design, construction, ownership, and operation and maintenance. The type of project supported by public-private partnerships varies from large-scale projects such as a waste treatment facility to small-scale community motor oil exchange programs.

Successful partnerships have certain elements in common, based on an EPA study of 23 partnerships:

- ◆ Local incentive to seek private assistance. For example, a community does not have the funds necessary to meet a 30 percent state recycling goal, thus local officials seek help from private companies.

- ◆ State legislation that gives local governments the authority to enter into private agreements.

- ◆ The private partner is assured a return on the investment.

- ◆ Small communities with limited revenue bases regionalize large-scale projects.

- ◆ Open communication between the private partner and the public.

- ◆ Agreement on the allocation of risks (e.g., financing, performance standards, and construction deadlines).

Source: EPA, *Public-Private Partnership Case Studies.*

NINE

WHAT CITIZENS CAN DO

"When politicians see enough people demanding action on the environment, then the laws will change."
—Vice-President Al Gore

Garbage is a national issue, but a local concern. While state and federal government set municipal solid waste management standards and requirements, overseeing and paying for the day-to-day management of waste continues to be the responsibility of local government. Paramount to the success of innovative solid waste programs is citizen involvement. Without citizen involvement, fewer community recycling programs would exist, fewer environmental education programs would be planned, fewer environmental workshops would be held, less legislative action would be taken, and the list goes on.

Citizens need not be experts on solid waste issues in order to make a contribution. Citizens like you add valuable perspectives to the dialogue and provide long-range views that may not be offered otherwise and insight into community values that technical experts may overlook. Facts and patience, not technical expertise, are the most important resources to effectively influence local, state, and federal policies.

Whether you are directly involved in the decision-making process, acting on behalf of a special or public interest group, or as a concerned citizen, consider the following steps to local action.

STEP ONE: GET THE FACTS

Find Out How Your Community Handles Its Municipal Solid Waste

A first step is to gather information about current solid waste management practices in your community. Sources for the information can include local departments of natural resources and public works, state and local environmental and health officials, local citizen and environmental groups, private waste disposal companies in the area, national environmental and public interest organizations, and the library. (FYI, some private companies may be reluctant to share what they consider "proprietary" information related to their costs and profits. However, persistence on your part can pay off). Identify the local experts on your issue.

Find answers to the following questions:

♦ Does your community have a comprehensive solid waste plan? Is it part of a state or regional plan?

♦ Do your community and state have a citizen advisory board or task force on solid waste?

♦ How does your community manage its municipal solid waste? How many tons per year are landfilled, incinerated, recycled, and/or composted?

♦ What are your community and state landfill, incinerator, and recycling regulations?

◆ Does your local solid waste facility have the required permits? Is it in compliance?

◆ When and how is the public notified if the landfill or incinerator fails to meet the requirements of Subtitle D, the Clean Air Act, the Clean Water Act, state laws, or local ordinances?

◆ What opportunities are there for the public to be informed and involved?

◆ What is the latest information on the different solid waste management options?

◆ What plans exist to meet future needs?

Find Out About Your Municipal Solid Waste Facilities

Operators of waste facilities are often the best source of information about the facilities themselves.

Ask:

◆ What is the capacity of the facility? In the case of landfills, how much landfill space remains and what is the projected closing date? In the case of incinerators, what are the contract requirements and what is the life of the plant?

◆ What pollution control equipment and procedures are used to protect human health and safety and the environment?

◆ What area is serviced by the facility and does the facility accept out-of-region or out-of-state waste?

◆ What plans exist to meet future needs?

◆ What is the facility's record for complying with laws and regulations relating to health, safety, and the environment? (For more information on waste facilities, see Chapters 3–5.)

Find Out About Your Garbage Disposal Rates

Rate information for municipal facilities and services for which the city or county contracts can be obtained from the local public works department. Rates for privately owned facilities and disposal services can be obtained directly from the company.

Ask:

- ◆ Do disposal and recycling rates reflect the full costs of MSW collection, facility operation and management, facility closure, and potential cleanup costs?

- ◆ How are disposal and recycling rates determined? Is there any plan to change the rates?

- ◆ How is the facility financed? Does it receive any state or federal funding?

- ◆ Are disposal and recycling costs expected to increase as a result of RCRA Subtitle D and other state and federal regulations? How will these costs be met?

Find Out How Consumers Pay For Waste Disposal Services

- ◆ Do consumers pay user fees? Are the fees visible to consumers (e.g., on their utility bill) or combined with other fees?

- ◆ Do local taxes include waste management services?

STEP TWO: WEIGH THE OPTIONS

Since municipal solid waste can be handled in a variety of ways, communities should carefully weigh the options available. Leaders and citizens alike should realize that any one waste management option may not meet the long-term needs of the community. What is needed is a comprehensive strategy for the present and the future. To begin assessing the alternatives, the following questions should be answered for each option:

Compare Waste Streams

- What kinds and amount of wastes will be handled?
- What is the amount by volume and weight of MSW that will be managed by each waste disposal option?

Compare Costs

- What are the full costs of the possible facilities: capital, operations and maintenance, closure and post-closure, and corrective action assurances?
- What financing options are available: ownership, bonds, taxes, user fees?

Compare Facility Operations And Management

- How will operating problems and emergencies be handled?
- What qualifications and training are required for facility employees?
- How and on what routes will waste be transported?
- How will excess waste and/or ash be handled?

Compare Economic Impacts

- What are the expected revenues?
- What are the expected costs?
- What are the projected economic impacts on the community and the possibility of compensation?
- Does the facility provide local employment opportunities?

Compare Environmental Impacts

- What are the potential environmental risks to groundwater, surface water, air, and land?
- What are the alternatives?

(For more information on the environmental impacts of waste facilities, see Chapters 3–5.)

Compare Social Impacts

- ◆ What are the potential impacts on the community other than environmental or economic?
- ◆ What is the level of public acceptance? Is the community divided? If so, why?
- ◆ What interest groups could play an active role? How and when should they be involved in the process?
- ◆ Do decision makers and the public have access to independent, unbiased advice?

Evaluate How To Manage The Project

Your community must decide if the facility or recycling program will serve the municipality or the region and whether it will be publicly or privately owned.

If your community decides to regionalize services, questions to ask should include:

- ◆ How will a regionally managed project affect the cost and oversight of the facility?
- ◆ What type of long-term oversight and monitoring will be provided for facility operations and environmental risks?
- ◆ Who will have access to environmental and financial data? Is an independent review board necessary?
- ◆ Does the community want to host a regional facility as a means to generate revenue or subsidize disposal costs?
- ◆ Is a regional facility more environmentally sound?

If your community decides to contract services to a private company, questions to ask should include:

- ◆ What arrangements will the community have for oversight of the facility?

- ◆ Will the community have access to company information? For example, is there access to the value and revenue of recyclables and the amount and type of materials handled?
- ◆ What training will be offered to the facility workers?
- ◆ How long is the contract? When and how is it to be reviewed and renewed?
- ◆ How is liability assigned?

STEP THREE: TAKE ACTION ON SOLID WASTE ISSUES

Build A Coalition

Any actions you decide to take will be more effective in coalition with other individuals and groups who share your goals. A coalition can pool resources, including information, political clout, funds, and volunteers. By drawing together others with similar concerns and interests, especially groups representing a cross-section of the community, a coalition also can demonstrate solid community support for change.

A coalition is only as strong as the groups it represents. Therefore, the more diverse the coalition, the more powerful its mandate. Diversity means representation by all the racial and socioeconomic groups affected by the solid waste issue. It is critical that each group's voice is not only heard, but that all viewpoints are included in the coalition's decision-making process. Everyone should work together to develop an agenda that reflects input from all of the coalition's member groups. Once the coalition's goals are identified, members should begin to promote its findings and goals to citizens and public officials in the community.

Take Your Message To The Community

Meet with a wide range of individuals to share your findings, your goals, and what it will take to meet these goals. Speak to individuals and groups that helped with your research, community leaders and decision makers, solid waste facility operators, public health and environmental officials, university professors, consumer advocates, and industry and business representatives.

Meet With Private Waste Companies Working In The Community

Discuss your goals and findings with private waste companies in the community, including waste haulers, facility owners, and recycling brokers.

Publicize Your Findings And Recommendations

Your coalition or group may want to publicize its findings. If that is the case, meet with local newspaper reporters and write letters to the editor reporting your findings. Urge local television and radio stations to cover the issues. Once the initial findings of the coalition are publicized, continue to keep the media informed of your progress and of upcoming meetings and events.

Take Legislative Action If Necessary

As you work on municipal solid waste issues, you may find that current rules and regulations no longer meet the needs of your community, or your coalition or group may develop alternative or new ideas and want to develop a regulatory climate that will help fulfill these goals. If such changes are needed you will need to present your case to the proper legislative body. For changes in land use management this may be the local zoning board; for environmental concerns it includes the local, county, and state environmental and natural resources departments and possibly the regional offices of the Environmental Protection Agency; for health concerns it includes the local, county, or state health department. In addition, you should enlist the support of state legislators and other elected officials. You also can call for and testify at public hearings. Finally, if the issue is important to your community, make it an election issue.

CONCLUSIONS

The construction of any solid waste disposal facility or the implementation of a solid waste management program will have an immediate as well as a long term impact on the community. Municipal solid waste management decisions must not be made hastily or simply address short-term needs. Solid waste management plans must assess the present and future needs of the area and ensure minimal risks to the community, the health and safety of the public, and the environment.

In the 1990s, attention has refocused on the federal government and its proper role in the management of the nation's trash. But whatever decisions are made at the federal level, waste management starts and ends in the community. Efforts to implement a solid waste plan or build a waste disposal facility remain in your hands. Too often, efforts to manage municipal solid waste result in a stalemate. But with accurate and understandable information, adequate technical support, and the opportunity to be heard, citizens can work toward solutions to our garbage problems.

Today, taking out the trash is everyone's chore.

NOTES

1. Environmental Protection Agency, *Characterization of Municipal Solid Waste in the United States: 1992 Update* (EPA/530-SW-90-042), p. ES-2.

2. Personal communication with Lisa Skumatz, Director of Synergic Resources Corporation. Survey conducted in 1991.

3. James E. McCarthy, "Solid Waste: RCRA Reauthorization Issues," *Congressional Research Service*, Updated March 3, 1992, p. 4.

4. Jim Glenn, "The State of Garbage in America: 1992 Nationwide Survey," *Biocycle*, April 1992, p. 50.

5. *Ibid.*, p. 54.

6. *Ibid.*

7. Jonathan Kiser, "Municipal Waste Combustion in North America: 1992 Update," *Waste Age*, November 1992, p. 30. Hospital figures from personal communication with Jim Edinger, EPA North Carolina Research Triangle Park.

8. David Minott, "Efficient Combustion With Fluid-Bed Furnaces," *Solid Waste & Power*, vol. IV, no. 5, October 1990, p. 36.

9. Congressional Research Service, *CRS Issue Brief: Solid Waste Management*, January 4, 1991, p. crs-3; Jim Glenn, "The State of Garbage in America," *Biocycle*, April 1992, p. 47.

10. Office of Technology, *Facing America's Trash: What Next for Municipal Solid Waste?*, 1989, p. 275.

11. NRDC, Sierra Club, The Izaak Walton League of America and Hazardous Waste Treatment Council, "Burning Used Oil—America's Undiscoverd Lead Threat," November 1991, p. i.

12. Books covering this issue include *Toxic Waste and Race in the United States,* Commission for Racial Justice, United Church of Christ, 1987; *Dumping in Dixie* by Robert Bullard, Westview Press, 1990; and *Minorities and the Environment: An exploration into the effects of environmental policies, practices and conditions on minority and low-income communities,* The Assembly, State of New York, Albany, 1991.

13. Mass burn WTE plant costs: EPA, *Decision-Makers Guide to Solid Waste Management,* 1989, p. 105; MRF costs: NSWMA, "The Cost To Recycle At a Materials Recovery Facility," p. 6; and personal communication with Chaz Miller, Manager Recycling Programs, NSWMA; Landfill costs: Robert T. Glebs, "Subtitle D: How It Will Effect Landfills?," *Waste Alternatives,* pp. 58–59.

▼

The timeline is adapted from National Solid Waste Management Association, "Garbage: Then & Now;" Martin Melosi, *Garbage in the Cities: Refuse, Reform, and the Environment 1880–1980, Resource Recycling, Waste Age;* and World Wildlife Fund, *Getting at the Source.*

RESOURCES

Publications/Videos

Beyond 40 Percent: Record Setting Recycling and Composting Programs. Institute for Local Self-Reliance, 2425 18th Street, NW, Washington, DC 20009, (202) 232-4108. 1991. $25. In a case study format, this study documents the operating experience of 17 communities, 14 of which have total materials recovery rates at or above 40 percent.

Burning Garbage in the US: Practice vs. State of the Art. Marjorie J. Clarke, Maarten de Kadt, Ph.D., and David Saphire. IN-FORM, Inc., 381 Park Avenue, New York, NY 10016, (212) 689-4040. 1991. $47. This book identifies the state of the art in incineration and provides an analysis of current incinerator performance.

"Changing Perspectives on the Facility Siting Process," *Maine Policy Review,* David Laws and Lawrence Susskind, Ph.D., December 1991, pp. 29–44. A review of the key steps in the siting process, alternative approaches, and its application in practice.

Characterization of Municipal Solid Waste in the United States: 1992 Update. Environmental Protection Agency, Solid Waste and Emergency Response, NTIS #PB92-207 166, (703) 487-4650. 1992. $19.50. For a free copy of the Executive Summary and Fact Sheet, EPA/530-s-92-019, (800) 424-9346. This report presents data from 1960 to 1990 on waste generation, disposal, and recovery.

Cleaning Up Toxics At Home and *Cleaning Up Toxics In Business.* California LWVEF. $29 for one video or $49 for both videos. Send orders to: The Video Project, 5332 College Ave., Suite 101, Oakland, CA 94618, (800) 4-Plant. These two 25-minute, award-winning videotapes creatively outline ways in which citizens and small business can significantly reduce pollution.

Clean From the Start: Composting Source-Separated Organics. Berkshire Film & Video, 33 Stormview Road, Lanesboro, MA 01237. 1992. $50. This video surveys 7 U.S. composting facilities and covers site design, odor control, marketing, and other issues.

Decision-Maker's Guide to Recycling Plastics. Oregon Department of Environmental Quality, Waste Reduction Section and EPA, Region X. Oregon DEQ, 811 S.W. Sixth Avenue, Portland, OR 97204, (503) 229-5913. 1990. This guide is to assist decision makers regarding whether and how to collect used plastics; also includes information on collection costs and recycling markets.

Decision-Makers Guide to Solid Waste Management, Volume I and Draft Volume II. Environmental Protection Agency, Solid Waste and Emergency Response, EPA/530-SW-90-019, (800) 424-9346. 1990. Free. Volume I is designed to help policy makers understand their present waste management problems. Volume II will contain more technical information on solid waste management.

Does the Solid Waste Management Hierarchy Make Sense? A Technical, Economic and Environmental Justification for the Priority of Source Reduction and Recycling. John Schall. Yale University Program on Solid Waste Policy, School of Forestry and Environmental Studies, 205 Prospect Street, New Haven, CT 06511-2189, (203) 432-3253. October 1992. $12. The first in a series of working papers on solid waste policy.

"Facility Siting Credo." National Workshop on Facility Siting sponsored by the Massachusetts Institute of Technology and the University of Pennsylvania's Wharton School of Business. 1990. A set of working guidelines for formulating more effective facility siting strategies distilled from the experience of dozens of siting experts and practitioners.

Facing America's Trash: What Next for Municipal Solid Waste? Office of Technology and Assessment, Superintendent of Documents, Government Printing Office, Washington, DC 20402, (202) 783-3238. 1989. $25. A discussion of options for a national policy based on the dual strategies of MSW prevention and better management.

The Garbage Dilemma: A Community Guide to Solid Waste Management. League of Women Voters of Illinois Education Fund, 332 South Michigan Avenue, Chicago, IL 60604, (312) 939-5935. 1990. $9. This guide on solid waste management options includes survey results of 81 Illinois communities' recycling efforts.

Garbage Solutions: A Public Officials Guide to Recycling and Alternative Solid Waste Management Technologies. Marian R. Chertow. U.S. Conference of Mayors and National Resource Recovery Association. 1620 I St., NW, Washington, DC 20006, (202) 293-7330. 1989. $14. This book provides a framework for decision making regarding recycling, source separation, composting, and special wastes.

Getting at the Source: Strategies for Reducing Municipal Solid Waste. World Wildlife Fund, Publications, P.O. Box 4866, Hampden Post Office, Baltimore, MD 21211, (410) 516-6951. 1990. $15. This report by the World Wildlife Fund and the Conservation Foundation offers detailed strategies for promoting source reduction.

The Green Report I and II: Recommendations for Responsible Environmental Advertising. Office of the Attorney General, 200 Ford Building, 117 University Avenue, St. Paul, MN 55155, (612) 246-7575. 1990. Free. A report by a 10-state task force of Attorneys General focusing on environmental marketing claims.

"Household Battery Recycling: Numerous Obstacles, Few Solutions," by Nancy Reutlinger and Dan de Grassi, *Resource Recycling,* April 1991, pp. 24-29. An article outlining the barriers obstructing battery recycling efforts, including the results of a 4-state survey of battery collection programs.

How to Deal with Controversy: Using Public Involvement in Successful Landfill Siting. Dayton Area League of Women Voters and American Association of University Women, Dayton, OH. Dayton Area LWV, 117 South Main Street, Suite 17, Dayton, OH 45402, (513) 228-4041. 1991. $1. This booklet provides a brief overview for public officials and includes successful landfill siting stories.

How to Set Up a Local Program to Recycle Used Oil. Environmental Protection Agency, Office of Solid Waste and Emergency Response, EPA/530-SW-89-039A, (800) 424-9346. May 1989. Free. A step-by-step manual outlining everything from the basic elements of a used-oil recycling program to designing and implementing the program, as well as promotion and maintenance.

How to Set Up an Environmental Shopping Program in Your Community. League of Women Voters of New Castle, P.O. Box 364, Chappaqua, NY 10514. $45 for a multimedia training kit to educate consumers, press, and local officials. This training kit shows how your organization can educate consumers, the press, and elected officials on how to reduce waste and toxics through purchasing decisions.

Incinerating Municipal Solid Waste: A Health Benefit Analysis of Controlling Emissions. Congressional Research Service, The Library of Congress. April 21, 1989. Available only through congressional offices. This report examines the potential health benefits of controlling emissions from MSW incinerators by considering 3 public policy strategies.

Is a Landfill Developer Looking At Your Community? Dayton Area League of Women Voters and American Association of University Women, Dayton, OH. Dayton Area LWV, 117 South Main Street, Suite 17, Dayton, OH 45402, (513) 228-4041. 1991. $1.50. This booklet provides background in local solid waste management issues and suggestions of ways to protect local interests if a landfill is proposed.

The Keystone Siting Process Handbook: A New Approach to Siting Hazardous and Nonhazardous Waste Management Facilities. Texas Water Commission, P.O. Box 13087, Austin, TX 78711. 1987. $5. This handbook, a revised edition of the Keystone Center's 1984 handbook, presents guidance for identifying and resolving issues associated with the siting of waste management facilities through the formation of a local review committee.

Keystone Siting Process Training Manual. Keystone Center, Box 606, Keystone, CO 80435, (303) 468-5822. 1985. $10. This manual, prepared for the League of Women Voters of Texas, is intended to be a user's guide and companion publication to the *Keystone Siting Process Handbook*. It provides a greater level of detail necessary for those who might become directly involved in the Keystone Siting Process.

Landfill Gas, Energy Utilization: Technology Options and Case Studies. Environmental Protection Agency, Office of Air and Radiation and Office of Policy, Planning and Evaluation. June 1992. Free. An assessment of the various landfill gas uses and limitations with case studies of 6 landfill gas energy projects in the U.S.

Making Less Garbage: A Planning Guide for Communities. INFORM, Inc., 381 Park Avenue, New York, NY 10016, (212) 689-4040. 1992. $30 ($15 for non-profits). This report documents successful government, business, and community source reduction strategies.

Markets for Scrap Tires. Environmental Protection Agency, Office of Solid Waste, EPA/530-SW-90-074A, (800) 424-9346. October 1991. Free. This report outlines the problems associated with scrap tires and identifies existing and potential source reduction and utilization methods.

Medical Waste Management in the United States: Second Interim Report to Congress. Environmental Protection Agency, Solid Waste and Emergency Response, EPA/530-SW-90-087A, (800) 424-9346. 1990. Free. A report regarding the tracking and management of medical wastes.

Mercury Rising: Government Ignores the Threat of Mercury from Municipal Waste Incinerators. Clean Water Action and Clean Water Fund, Publications, 1320 18th Street, NW, Washington, DC 20036, (202) 457-1286. 1990. Free. Documents the growing threat of mercury emissions from municipal waste incinerators.

Plastic Waste Primer: A Handbook for Citizens. League of Women Voters Education Fund, Publications, 1730 M Street, NW, Washington, DC 20036, (202) 429-1965. Pub. #954, 1992. $10.95 ($8.95 for members). This report provides information to help concerned citizens and local government leaders understand the issues, evaluate the arguments, and make decisions about plastics in the MSW stream.

Proposed Dioxins Control Measure for Medical Waste Incinerators: Staff Report. State of California Air Resources Board, Stationary Source Division. Public Information Office, 1102 Q Street, Sacramento, CA 95814, (916) 322-2990. 1990. A presentation of recommendations designed to reduce dioxin emissions from medical waste incinerators as well as a cost/benefit analysis of options.

Public-Private Partnership Case Studies: Profiles of Success in Providing Environmental Services. Environmental Protection Agency, Administration and Resources Management, EPA/20M-2005, (202) 245-4020. 1990. Free. This report provides examples of how partnerships work and are developed, including lessons learned and useful information on developing or choosing partnerships.

Recycling & Incineration: Evaluating the Choices. Richard Denison and John Ruston. Island Press, (800) 828-1302. 1990. $19.95. This book examines the technology, economics, environmental concerns, and legal intricacies behind these 2 approaches for solid waste management.

Recycling Is More Than Collections: Questions & Concerns from the Ground Up. League of Women Voters Education Fund, Publications, 1730 M Street, NW, Washington, DC 20036, (202) 429-1965. Pub. #926, 1991. $5.95 ($4.95 for members) plus shipping and handling. An examination of the economic and policy challenges to expanding recycling markets and a report on the findings of a nationwide survey of local solid waste officials and newspaper publishers.

Recycling Works! State and Local Solutions to Solid Waste Management Problems. Environmental Protection Agency, Solid Waste and Emergency Response, EPA/530-SW-89-014, (800) 424-9346. January 1989. Free. This booklet provides information about successful recycling programs initiated by state and local agencies, including private and joint recycling ventures.

The Road to Less Waste: Recycling New York State's Scrap Tires Into Asphalt Paving Material. A staff report to the Chairman, Assembly member Maurice D. Hinchey, NYS Legislative Commission on Solid Waste Management, Empire State Plaza, Agency Building #4, 5th Floor, Albany, NY 12248, (518) 455-3711. January 1991. This report provides an overview of the scrap tire problem and its use in asphalt paving material.

Sites for Our Solid Waste: A Guidebook for Effective Public Involvement. Environmental Protection Agency, Solid Waste and Emergency Response, EPA/530-SW-90-019, (800) 424-9346. 1990. Free. This policy-makers guidebook presents a strategy for effectively resolving conflicts that might appear during the siting process.

Siting Hazardous Waste Management Facilities: A Handbook. Conservation Foundation, National Audubon Society and Chemical Manufacturers Association. National Audubon Society, 115 Indian Mound Trail, Travernier, FL 33070. $2. Questions and information to help individuals decide whether a particular hazardous waste management facility should be built at a given site.

Solid Waste Landfills. University of Wisconsin and Waste Age. Philip O'Leary, Engineering Professional Development, University of Wisconsin-Madison, 432 North Lake Street, Madison, WI 53706, (608) 262-0493. 1986. An 8-part lesson on the design and operation of sanitary landfills.

Solid Waste: RCRA Reauthorization Issues. Congressional Research Service, The Library of Congress. Updated monthly. Available only through congressional offices. A summary of the key legislative issues in the reauthorization of RCRA.

State of the World 1991: A Worldwatch Institute Report on Progress Toward a Sustainable Society. W.W. Norton & Company. $10.95. An assessment of global issues pertaining to the health of the planet, with a chapter on "Reducing Waste, Saving Materials."

State Recycling Laws Update. Raymond Communications, 6429 Auburn Avenue, Riverdale, MD 20737, (301) 345-4237. February 1993. $115. A summary of state recycling legislation and analysis of key recycling issues.

Toxic Wastes and Race in the United States. Commission for Racial Justice, United Church of Christ, 700 Prospect Avenue, Cleveland, OH 44115-1110. 1987. $20. A national report on the racial and socioeconomic characteristics of communities with hazardous waste sites.

Trade-offs Involved in Beverage Container Deposit Legislation. General Accounting Office, P.O. Box 6015, Gaithersburg, MD 20877, (202) 275-6241. November 1990. Free. An examination of beverage container deposit laws: their effects and the level of public support including their compatibility with curbside recycling programs.

"The Unsavory Side of Battery Recycling," by Tom Watson, *Resource Recycling,* April 1991, pp. 46-50. An article regarding a West Coast battery recycler charged with waste management violations.

What Do the Standards Mean? A Citizens' Guide to Drinking Water Contaminants. Carolyn J. Kroehler. Virginia Water Resources Research Center, 617 N. Main Street, Blacksburg, VA 24060. Free to Virginia residents, out-of-state $8. A guide on drinking waters standards and contaminants, including treatment options and a useful cross-reference list of contaminants.

Variable Rates in Solid Waste: Handbook for Solid Waste Officials, Volumes I and II. Lisa Skumatz, Ph.D. and Cabell Breckinridge. Environmental Protection Agency, Region 10, Solid Waste Program, HW-072, 1200 6th Avenue, Seattle, WA 98101, (206) 442-6641. June 1990. Volume I $10 and Volume II $37.50. A detailed guide on how to initiate a pay-per-can garbage collection system, including case studies on Seattle and other cities.

Periodicals/Magazines

Biocycle: Journal of Waste Recycling. J.G. Press, Inc., Box 351, 419 State Avenue, Emmaus, PA 18049, (215) 967-4135. $55/year, 12 issues. A monthly journal specializing in organic composting, recycling, and reuse.

E: The Environmental Magazine. Earth Action Network, Inc., 28 Knight Street, Norwalk, CT 06851, (203) 854-5559. $20/year, 6 issues. A bimonthly magazine covering a wide range of environmental issues with regular articles on solid waste management.

EHMI Re:Source. Environmental Hazards Management Institute, 10 Newmarket Road, P.O.Box 70, Durham, NH 03824, (603) 868-1496. A quarterly newsletter providing information on environmental management.

Environmental Action Magazine. 6930 Carroll Avenue, Suite 600, Takoma Park, MD 20912, (301)-891-1100. 1525 New Hampshire Avenue, NW, Washington, DC 20036, (202) 745-4870. $20 membership, bimonthly magazine. An environmental publication by the organization Environmental Action. This bimonthly magazine focuses on toxics, solid waste and energy issues, and legislation.

EPA Reusable News. Environmental Protection Agency. Office of Solid Waste and Emergency Response, OS-305, 401 M Street, SW, Washington, DC 20460. A quarterly newsletter on EPA's efforts and others regarding MSW management.

EPA Used Oil Recycling. Environmental Protection Agency. Office of Solid Waste and Emergency Response, OS-323, 401 M Street, SW, Washington, DC 20460. A quarterly newsletter covering local, state, and national news and developments regarding used oil recycling.

Garbage: The Practical Journal for the Environment. Old House Journal Corp., 435 9th Street, Brooklyn, NY 11215, (718) 788-1700. $21/year,

12 issues. An environmental magazine covering topical recycling and waste management issues.

Household Hazardous Waste Management News. Waste Watch Center, 16 Haverhill Street, Andover, MA 01810, (508) 470-3044. Published quarterly; subscriptions are free. A newsletter focusing on regional, national, and international household hazardous waste management efforts.

Recycling Times. Waste Age's Recycling Times, 5615 W. Cermak Road, Cierco, IL 60650. $95/year, 26 issues. A biweekly newspaper on recycling markets published by the National Solid Waste Management Association.

Recycling World. Environmental Defense Fund, 257 Park Avenue South, New York, NY 10010. Published irregularly; single copy free with stamped, self-addressed envelope. A newsletter promoting practical action for the environment.

Resource Recycling. P.O. Box 10540, Portland, OR 97210-9893, (800) 227-1424. $42/year, 12 issues. A monthly magazine covering national and local recycling efforts.

Scrap Tire News. Recycling Research Institute, 133 Mountain Road, Suffield, CT 06078 (203) 668-5422 or 8727 Beechwood Drive, Fairfax, VA 22031 (703) 280-9112. $118/year. A monthly newsletter covering news and developments in the scrap tire processing industry.

Solid Waste & Power: The Magazine of Waste Management Solutions. HCI Publications, 410 Archibald Street, Kansas City, MO 64111-3046, (816) 931-1311. $49/year, 7 issues. A bimonthly magazine aimed at WTE systems and recycling.

Waste Age. 1730 Rhode Island Avenue, Suite 1000, NW, Washington, DC 20036, (202) 861-0708. $45/year, 12 issues. A monthly magazine focusing on the industry and technology of waste systems.

Wastelines. Environmental Action Foundation, 6930 Carroll Avenue, Suite 600, Takoma Park, MD 20912, (301) 891-1100. $10/year. A quarterly environmental newsletter geared toward the citizen activist. Focus is at the state level but also covers important federal legislation.

World Wastes: The Independent Voice. Fulfillment Department, P.O. Box 41369, Nashville, TN 37204-1094, (615) 377-3322. $45/year, 12 issues. A monthly magazine covering a wide range of waste management systems.

Organizations

American Plastics Council. 1275 K Street, NW, Suite 400, Washington, DC 20005, (202) 371-5319. A plastics industry trade group promoting plastics recycling and research. APC operates a toll-free number to answer general questions about plastics products, plastics recycling opportunities, and potential markets for collected plastics: (800) 243-5790.

Californians Against Waste Foundation. 926 J Street, Suite 606, Sacramento, CA 95814, (916) 443-8317. CAW is dedicated to creating a sustainable economy through the reduction, reuse, recycling, and composting of our resources.

Citizen's Clearinghouse for Hazardous Wastes. P.O. Box 6806, Falls Church, VA 22040, (703) 237-CCHW. This organization is concerned with the physical effects of contact with toxic chemicals and other hazardous wastes. Conducts site visits to determine the severity of toxic pollution and conducts research on chemicals to determine dangerous levels of usage.

Coalition of Northeastern Governors Policy Research Center, Inc. 400 North Capitol Street, NW, Washington, DC 20001, (202) 624-8450. This organization has a special task force on source reduction that is pressing for legislation to reduce the amount of materials used in packaging.

Composting Council (formerly the Solid Waste Composting Council). 114 South Pitt Street, Alexandria, VA 22314, (703) 739-2401 or (800) 457-4474. The Council's principal objectives are to promote composting, encourage the production of high quality compost, define compost as recycled material, and serve as an information clearinghouse.

Conservation Law Foundation. 3 Joy Street, Boston, MA 02108, (617) 742-2540. CLF uses law to improve resource management, environmental protection, and public health throughout New England.

Environmental Action Foundation. 6930 Carroll Avenue, Suite 600, Takoma Park, MD 20912, (301) 891-1100. An environmental organization, EAF offers technical assistance and an information clearinghouse for groups interested in recycling programs. Publishes a newsletter called *Wasteline*.

Environmental Defense Fund. 275 Park Avenue South, New York, New York 10010, (212) 505-2100. A national environmental organization,

EDF conducts research on recycling, source reduction, and incineration.

Environmental Research Foundation. P.O. Box 73700, Washington, DC 20056-3700, (202) 328-1119. Provides technical assistance to grassroots environmental groups and publishes a weekly bulletin, *Rachel's Hazardous Waste News.*

INFORM, Inc. 381 Park Avenue South, New York, NY 10016, (212) 689-4040. A nonprofit research and public education organization that identifies and reports on practical actions for the conservation and preservation of natural resources and public health.

Institute for Local Self-Reliance. 2425 18th Street, NW, Washington, DC 20009, (202) 232-4108. A nonprofit research organization that offers technical assistance to municipal governments and community groups to develop more efficient, self-contained waste management systems.

Integrated Waste Services Association. Two Lafayette Centre, 1133 21st Street, NW, Washington, DC 20036, (202) 467-6240. A trade association representing the WTE industry.

Keystone Center. Box 606, Keystone, CO 80435, (303) 468-5822. A nonprofit national center for environmental negotiation, training, and education.

National Audubon Society. 666 Pennsylvania Avenue, SE, Washington, DC 20003, (202) 547-9009. An environmental organization concerned with the long-term protection and use of land, water, wildlife, and other natural resources.

National League of Cities. 1301 Pennsylvania Avenue, NW, Suite 600, Washington, DC 20004, (202) 626-3000. An organization representing the interests of municipalities at the federal level.

National Recycling Coalition. 1101 30th St., NW, Suite 305, Washington, DC 20007, (202) 625-6406. A coalition of recycling industries and nonprofit groups that distributes publications and publication lists on recycling, lists state recycling contacts, holds an annual conference, and sponsors annual recycling awards.

National Solid Wastes Management Association. 1730 Rhode Island Avenue, NW, Suite 1000, Washington, DC 20036, (202) 659-4613. A trade group representing the waste services industry. Collects and analyzes data on numerous aspects of waste disposal and recycling.

Natural Resources Defense Council. 122 East 42nd Street, New York, NY 10168, (212) 949-0049. NRDC is dedicated to protecting endangered natural resources by combining legal and scientific approaches to monitor government agencies, bring legal action, and disseminate citizen information.

New Hampshire Resource Recovery Association. P.O. Box 721, Concord, NH 03302, (603) 224-6996. This association provides information and technical assistance to communities to develop recycling market cooperatives.

Project ROSE (Recycling Oil Saves Energy). The University of Alabama, Box 870203, Tuscaloosa, AL 35487-0203, (800) 452-5901 (in Alabama), (205) 348-4878 (outside Alabama). This organization collects used oil from individual, corporate, and municipal consumers, garages, and service stations for treatment by a used-oil processor.

Sierra Club. National Headquarters, 730 Polk St., San Francisco, CA 94109 or Washington Office, 408 C Street, NE, Washington, DC 20002, (202) 547-1141. A grassroots environmental organization dedicated to promoting the responsible use of the earth's ecosystems and resources through advocacy and legislative action.

Solid Waste Association of North America. P.O. Box 7219, Silver Spring, MD 20910, (301) 585-2898. An association primarily serving public sector officials responsible for managing and operating MSW management systems.

Southwest Research and Information Center. Box 4524, Albuquerque, NM 87106, (505) 262-1862. A community-oriented, nonprofit educational and scientific organization providing assistance on water quality, toxic wastes, and a variety of other environmental issues at the local, state, and federal levels.

United Church of Christ, Commission for Racial Justice. 475 Riverside Drive, New York, NY 10115, (212) 870-2162. The Commission is a national church-based civil rights agency committed to racial justice and reconciliation.

U.S. Conference of Mayors. 1620 I Street, NW, Washington, DC 20006, (202) 293-7330. This organization represents the nation's cities with populations of 30,000 or more on a broad range of issues facing urban America.

United States Public Interest Research Group (U.S. PIRG). 215 Pennsylvania Avenue, SE, Washington, DC 20003, (202) 546-9707. A public interest group engaged in research, investigation, and legislative action.

Wisconsin Department of Natural Resources. SW3, P.O. Box 7921, Madison, WI 53707. This government office provides information and technical assistance to communities to develop recycling market cooperatives.

World Wildlife Fund/Conservation Foundation. 1250 24th Street, NW, Washington, DC 20037, (202) 293-4800. This organization works to preserve the health of ecological systems and promote sustainable use of natural resources, more efficient resource and energy use, and reduction of pollution.

State and Local Leagues of Women Voters. Every state and Puerto Rico, the District of Columbia, and the Virgin Islands have a state League and one or more local Leagues. Check your phone directory for a listing.

Government Assistance and Clearinghouses

RCRA/Superfund Hotline: (800) 424-9346. Provides information and compliance requirements.

Solid Waste Assistance Program: (800) 677-9424. Provides information and general technical assistance on integrated MSW management.

Recycled Products Information Clearinghouse: (703) 941-4452. 5528 Hempstead Way, Springfield, VA 22151. Provides recycled product and market information.

U.S. EPA Regional Offices: Region I, Boston, MA; Region II, New York, NY; Region III, Philadelphia, PA; Region IV, Atlanta, GA; Region V, Chicago, IL; Region VI, Dallas, TX; Region VII, Kansas City, MO; Region VIII, Denver, CO; Region IX, San Francisco, CA; Region X, Seattle, WA

INDEX